# Synthesis Lectures on Power Electronics

**Series Editor**

Jerry Hudgins, Lincoln, USA

This series publishes short books on topics related to power electronics, ancillary components, packaging and integration, electric machines and their drive systems, as well as related subjects such as EMI and power quality. Each Lecture develops a particular topic with the requisite introductory material and progresses to more advanced subject matter such that a comprehensive body of knowledge is encompassed. Simulation and modeling techniques and examples are included where applicable.

Yonglu Liu

# Active Power Decoupling Technology in Single-Phase Current-Source Converters

 Springer

Yonglu Liu (iD)
School of Automation
Central South University
Changsha, China

ISSN 1931-9525                    ISSN 1931-9533  (electronic)
Synthesis Lectures on Power Electronics
ISBN 978-3-031-21272-7          ISBN 978-3-031-21270-3  (eBook)
https://doi.org/10.1007/978-3-031-21270-3

This Springer imprint is published by the registered company Springer Nature Switzerland AG
The registered company address is: Gewerbestrasse 11, 6330 Cham, Switzerland

# Preface

Single-phase power electronics interface is widely used in various occasions, such as Photovoltaic (PV) power generation, Light-Emitting Diode (LED) lighting driving, and electric vehicle charging. It is well known that the instantaneous power at the AC side includes a second-order ripple power. Nowadays, the second-order ripple power is usually handled by employing bulky passive components (called passive power decoupling technology). This method is easy to implement, but large volume and weight are undesirable from the cost and practicability perspective. Moreover, the life expectancy of large aluminum electrolytic capacitors is limited. Then, active power decoupling technology has been proposed and developed recently. It removes usage of bulky passive components and enhances power density and reliability. Undoubtedly, the active power decoupling technology meets the development trends of power electronic devices.

This book gives a comprehensive and in-depth introduction into active power decoupling technology and its most recent topics. The discussion will be centered mainly around the single-phase current-source converter. It presents topology construction principles, specific decoupling topologies, advanced decoupling controls, and stability analysis methods. Generally, the introduced topology construction principles and the control concepts can also be extended to other converters, such as single-phase voltage-source converters, three-phase converters, and cascaded multilevel converters. With this book, I intend to enable the reader to get the latest knowledge on the concept and operation of the active power decoupling technology from the fundamental to the whole picture and help researchers, engineers, and designers to properly design or select the decoupling circuit topology and the control strategy according to the specific application.

This book consists of eight chapters based on several research projects, covering the aspects of converter topologies and power control strategies for active power decoupling converters. It starts with an introduction in Chap. 1 to present a brief introduction focusing on the applications of single-phase power electronics, 2-order ripple power issue, passive power decoupling technique, and active power decoupling technique. For the rest contents, from the view of circuit topologies and control strategies, this book can be divided into two parts. The first part is focused on topology structures which are from Chaps. 2–5. And the second part presents the control strategies and stability analysis, which are included from Chaps. 6–8. The detail organization of the book content is summarized in Chap. 1.

First, I want to thank my main supervisor Prof. Yao Sun for giving me the opportunity to do my Ph.D. thesis in the field of power electronics. I sincerely appreciate all his contribution of time and numerous ideas to make my Ph.D. experience productive as well as stimulating. I also want to express my sincere appreciation to Prof. Mei Su and Prof. Hua Han for helping me to be a better person, both personally and professionally. Many thanks to all my colleagues for their support, mentorship, and friendship. Although it is not a complete list, I will mention some of those friends who provided valuable input and constructive criticism to my work directly or indirectly. They are Hui Wang, Hanbing Dan, Wenjing Xiong, Guo Xu, Guangfu Ning, Zhangjie Liu, Jianheng Lin, and so many others.

Then, I want to express my sincere gratitude to my parents, my parents-in-law, my wife, my sister, and my daughter for their support and for keeping me motivated.

Finally, the authors are extremely grateful to Springer and the editorial staff for the opportunity to publish this book and for help in all possible manners.

Changsha, China                                                                      Yonglu Liu

# Contents

# List of Figures

# Introduction

Single-phase converter, as one of basic power electronic converters, has been widely used in industrial, commercial and residential fields. The specific applications contain light-emitting-diodes (LEDs) drivers, photovoltaic (PV) power generation, fuel cell (FC) power generation, electric vehicle (EV) battery chargers, railroad traction power supply, uninterruptible power supply (UPS), power electronic transformers, cascaded multilevel converters and so on. However, the 2-order ripple power inherently exists in single-phase converters, which brings detrimental effects to the systems. Addressing this issue, the solutions include the passive and active power decoupling methods. The former is widely used nowadays and the latter is the development trend to achieve high power density. In this chapter, the single-phase system and its applications are firstly introduced in Sect. 1.1. Then, Sect. 1.2 analyzes the 2-order power issues. The passive and active power decoupling methods are presented in Sects. 1.3 and 1.4. Section 1.5 provides a brief overview of the contents and organization of the entire book.

## 1.1    Single-Phase System and Its Applications

The single-phase converter is the electrical energy conversion interface between AC and DC power. In terms of function, it can be divided into rectifiers (AC-DC converters) and inverters (DC-AC converters). Four basic single-phase rectifiers and inverters are demonstrated in Fig. 1.1. Figure 1.1a is the single-phase full-bridge voltage-source rectifier/inverter. Figure 1.1b is the widely used single-phase diode rectifier with a boost power factor correction (PFC) circuit. It provides a stiff DC bus voltage with less active switches. However, it only can operate under unity power factor. Figure 1.1c, d are the single-phase full-bridge current-source rectifier and its inversion version. The DC bus is

Y. Liu, *Active Power Decoupling Technology in Single-Phase Current-Source Converters*, Synthesis Lectures on Power Electronics, https://doi.org/10.1007/978-3-031-21270-3_1

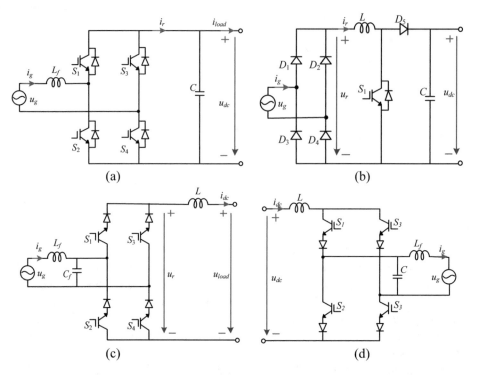

**Fig. 1.1** Single-phase current-source and voltage-source converters. **a** Single-phase full-bridge voltage-source rectifier/inverter. **b** Single-phase diode rectifier with PFC. **c** Single-phase full-bridge current-source rectifier. **d** Single-phase full-bridge current-source inverter

a stiff current. The following is a brief introduction to some typical applications of these single-phase converters.

(1) *LEDs Driver*

The global lighting consuming takes about 20% of the total electricity generation. Therefore, it is important to develop the green and energy-efficient lighting technique. LED lights have the advantages of high luminous efficiency, long lifespan, zero pollution, etc., thus being widely used in indoor and outdoor lighting, traffic signals, decorative lights and so on. Different from incandescent lamps and fluorescent lamps, LED lighting devices are driven by a DC current source. Therefore, AC-DC power conversion supplies are required. Figure 1.2 depicts the structure diagram of an AC-DC LED driver. It should have the characteristic of high efficiency, long lifespan, and constant output current to fully utilize the advantages of LEDs.

**Fig. 1.2** Structure diagram of
an AC-DC LED driver

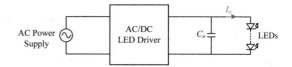

(2) *PV Power Generation System*

Among various renewable energy sources, PV power generation has shown a rapid development in recent decades due to the advantages of abundant resources, wide distribution
and simple installation. Its cost is significantly reduced with the progress of relative tech-
nologies. According to the European Commission's Joint Research Center (JRC), PV
power generation will account for 5% of the world's total power generation by the middle of this century, and the proportion will rise to 64% by the end of this century. The
development of PV power generation is an important cornerstone to achieve the goal of
"carbon neutrality".

PV power generation systems can be divided into grid-connected and grid-off systems.
The grid-connected system provides the power system with active and reactive power. The
grid-off system is not connected to the grid and serves as a mobile power source to supply
remote areas without electricity. As shown in Fig. 1.3, the single-phase DC-AC inverter is
used to connect the PV cell with the grid or the user loads. Specifically, PV grid-connected
inverters can be classified into isolated and non-isolated inverters depending on whether
there are transformers or not. The selection and design of a PV inverter is very important
because critical performance metrics like efficiency, cost, power density, leakage current
and reliability, are directly affected.

(3) *FC Power System*

FC directly converts the fuel chemical energy into electrical energy through electrochemical reactions. It possesses high conversion efficiency since the electrochemical reaction

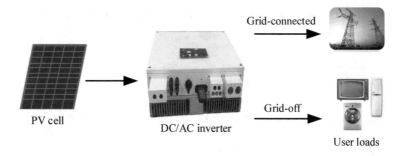

**Fig. 1.3** Structure of PV power generation system

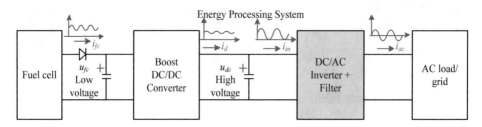

**Fig. 1.4** Structure diagram of the fuel cell power supply system

is not limited by the Carnot cycle. With the advantages of almost zero emission, convenience of installation and immunity to environment interference, FC power system has been widely applied to spacecraft, space shuttle, submarine power sources, and portable power supplies.

Figure 1.4 shows the structure of the fuel cell system, which is essentially a DC-AC power conversion system. It contains the FC, a boost DC-DC converter, a DC-AC inverter and the AC load/grid. If there exits low-frequency ripple current in $i_{fc}$, the hysteresis will be caused in the fuel cell, which will threaten the safe operation. Besides, the fuel utilization and the power supply efficiency will deteriorate. Therefore, the used converters and control strategies are critical to suppress or remove the low-frequency current ripple.

(4) *EV Battery Charger*

The energy consumption and the exhaust emission of vehicles make the automotive industry develop towards electrification, low carbonization and intelligence. Developing EVs is an effective way to solve the problems of global energy shortage and environmental pollution. Then, chargers are needed to transfer the power from the gird to the EV battery. Figure 1.5 shows the schematic of EV charging. The charging piles can be divided into the AC charging pile and the DC charging pile. The DC charging pile directly supplies to vehicle batteries, while on-board chargers are needed for the AC charging pile to convert the AC power to DC power. The DC charging pile and on-board charger are AC-DC converters, which have significant impacts on the charging efficiency and the battery lifetime.

## 1.2    2-Order Ripple Power Issue

For the single-phase system, assume the grid voltage $u_g$ and current $i_g$ are expressed as:

$$u_g = V \cos(\omega t) \tag{1.1}$$

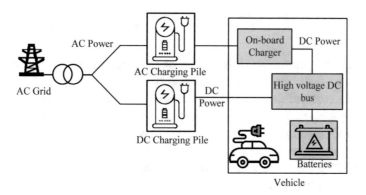

**Fig. 1.5** Schematic of EV charging

$$i_g = I \cos(\omega t + \varphi) \tag{1.2}$$

where $V$ and $I$ are the amplitudes of the gird voltage and current, $\omega$ is the grid frequency and $\varphi$ is the displacement angle. Then the instantaneous power $p_{ac}$ at the grid side is

$$p_{ac} = \underbrace{\tfrac{1}{2}VI \cos(\varphi)}_{P_o} + \underbrace{\tfrac{1}{2}VI \cos(2\omega t + \varphi)}_{P_r} \tag{1.3}$$

where $P_o$ is the average power (i.e., the load power) and $P_r$ is the ripple power, which is a 2-order component. The power composition of the single-phase system is shown in Fig. 1.6.

For simplicity, the effects of power losses and input filters are neglected. According to power balance, the rectified output voltage $u_r$ of the current-source converter (CSC) and output current $i_r$ of the voltage-source converter (VSC) are expressed as:

$$u_r = u_g i_g / i_{dc} = \underbrace{0.5VI\cos(\varphi)/i_{dc}}_{u_{load}} + \underbrace{0.5VI\cos(2\omega t + \varphi)/i_{dc}}_{\tilde{u}_r} \tag{1.4}$$

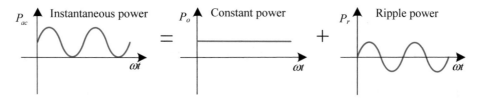

**Fig. 1.6** Single-phase system power composition

$$i_r = u_g i_g / u_{dc} = \underbrace{0.5VI\cos(\varphi)/u_{dc}}_{i_{load}} + \underbrace{0.5VI\cos(2\omega t + \varphi)/u_{dc}}_{\tilde{i}_r} \qquad (1.5)$$

where $i_{dc}$ and $u_{dc}$ are the DC-link current of CSC and the DC-link voltage of VSC, respectively; $\tilde{u}_r$ and $\tilde{i}_r$ are the ripple components.

The ripple components cause adverse effects to the energy conversion systems. In LED lighting system, the low-frequency flicker will happen, which affects the lighting quality. In PV power generation system, the 2-order ripple current $\tilde{i}_r$ leads to power oscillation near the maximum power point, which hinders the maximum power point tracking (MPPT) and reduces the system efficiency. In FC system, the low-frequency harmonic components in output current will cause voltage-current hysteresis phenomenon, which shortens the fuel cell life, and reduces the fuel efficiency and output capacity. Therefore, the 2-order ripple power must be handled to alleviate or even eliminate the above-mentioned adverse effects.

## 1.3    Passive Power Decoupling

To solve the 2-order ripple power issue, the traditional approach is to increase the size of the passive components, which is called the passive decoupling method. It includes employing bulky capacitors/inductors or an LC resonant circuit. The corresponding working principles are respectively demonstrated as follows.

(1) *Increasing the Size of the Capacitors/Inductors*

Referring to Fig. 1.1a, c, taking a resistor load as an example, the dynamic equations of the DC-link are

$$L\frac{di_{dc}}{dt} = u_r - Ri_{dc} \qquad (1.6)$$

$$C\frac{du_{dc}}{dt} = i_r - \frac{u_{dc}}{R} \qquad (1.7)$$

For (1.6), considering the initial condition $i_{dc}(t = 0) = 0$, we have the solution as follows,

$$\begin{cases} i_{dc}(t) = \sqrt{a - b} \\ a = 0.5VI\dfrac{R^2\cos(2\omega t + \varphi) + R^2\cos(\varphi) + L^2\omega^2\cos(\varphi) + LR\omega\sin(2\omega t + \varphi)}{R(R^2 + L^2\omega^2)} \\ b = 0.5VI\dfrac{L^2\omega^2\cos(\varphi) + LR\omega\sin(\varphi) + 2\cos(\varphi)R^2}{e^{2Rt/L}R(R^2 + L^2\omega^2)} \end{cases} \qquad (1.8)$$

In (1.8), $a$ is the steady term and $b$ is the transient term. Finally, $b$ will be zero and the steady DC current $i_{dc}$ is

$$
\begin{cases}
i_{dc}(t) = \sqrt{0.5VI\left(\dfrac{\cos(\varphi)}{R} + \dfrac{\sin(2\omega t + \varphi + \theta)}{\sqrt{R^2 + L^2\omega^2}}\right)} \\[4mm]
\tan\theta = \dfrac{R}{\omega L}
\end{cases}
\tag{1.9}
$$

Similarly, according to (1.5) and (1.7), the dc bus voltage $u_{dc}$ can be expressed as

$$
\begin{cases}
u_{dc}(t) = \sqrt{c - d} \\[2mm]
c = 0.5VIR\dfrac{\cos\varphi\left(\omega^2 C^2 R^2 + 1\right) + \omega CR\sin(2\omega t + \varphi) + \cos(2\omega t + \varphi)}{\left(\omega^2 C^2 R^2 + 1\right)} \\[4mm]
d = 0.5IR\dfrac{\omega CR\sin\varphi + V\cos\varphi + V\cos(\varphi)\left(\omega^2 C^2 R^2 + 1\right)}{e^{2t/CR}\left(\omega^2 C^2 R^2 + 1\right)}
\end{cases}
\tag{1.10}
$$

And the steady DC bus voltage $u_{dc}$ is

$$
\begin{cases}
u_{dc}(t) = \sqrt{0.5VIR\left(\cos\varphi + \dfrac{\sin(2\omega t + \varphi + \alpha)}{\sqrt{\omega^2 C^2 R^2 + 1}}\right)} \\[4mm]
\tan\alpha = 1/\omega CR
\end{cases}
\tag{1.11}
$$

From (1.9) and (1.11), there exists low frequency components in the DC current/voltage. And they reduce with increasing the size of $L/C$ (making the term $1/\sqrt{R^2 + L^2\omega^2}$ or $1/\sqrt{\omega^2 C^2 R^2 + 1}$ sufficiently small). However, to realize perfect constant DC current/voltage, the value of $L/C$ needs to be infinite.

(2) *Adding an LC Resonant Circuit*

Figure 1.7a illustrates a CSC with a parallel LC resonant circuit, which is in series with the DC-link. In order to achieve 2-order ripple power decoupling, the resonant frequency is designed as

$$
f_r = \frac{1}{2\pi\sqrt{L_d C_d}} = 2f_{ac}
\tag{1.12}
$$

where $f_{ac}$ is the grid frequency. For the 2-order ripple voltage ($\tilde{u}_r$), the impedance of the resonant circuit is infinite and $u_{ab}$ will be equal to $\tilde{u}_r$. Then, no low frequency voltage is imposed on the DC bus inductor and its size can be small.

In terms of the VSC, the adopted series LC resonant circuit is in parallel with the DC-link as shown in Fig. 1.7b. The resonant frequency is also the twice the grid frequency.

**Fig. 1.7** Single-phase
converters with an LC resonant
circuit in **a** CSC and **b** VSC

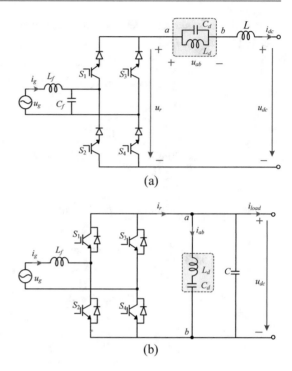

(a)

(b)

For the 2-order ripple current $(\tilde{i}_r)$, the impedance of the resonant circuit is zero and $i_{ab}$ will be equal to $\tilde{i}_r$.

However, the required resonant inductance and capacitance are still large due to the relatively low resonant frequency, i.e., 100 Hz or 120 Hz. In addition, with the parameter drifts, the decoupling effect deteriorates and a circulating current will generate to pose a risk of overvoltage/overcurrent.

For the passive power decoupling technique, no extra semiconductor devices are added and the circuit structure is relatively simple. The obvious disadvantage is the requirement of bulky passive elements. The bulky inductor reduces the power density significantly and slows the dynamic response. In addition, 2-order ripple current increases core losses. For the large aluminum electrolytic capacitor (it is usually used when the needed capacity is large), another drawback is the limited lifespan.

## 1.4  Active Power Decoupling

The active power decoupling technology uses switching devices to divert the 2-order ripple power from the main circuit to a small energy storage element. The ripple power is buffered by swinging the voltage/current with a large fluctuation. Then, no large inductor or capacitor is used and system reliability and power density are enhanced. So far as the research progressed, its application scenarios have expanded from low-voltage low-power converters to high-voltage high-power cascaded multilevel converters. Besides, the functions of decoupling circuits are no longer limited to power decoupling. And other functions, such as multilevel implementation, power factor correction, leakage current suppression, and output voltage range extension, begin to be integrated. This injects new vitality into the active power decoupling technology. At present, active power decoupling technology has become a hot spot in the research field of power electronics.

### 1.4.1 Working Principle

The idea of the active power decoupling technology is demonstrated in Fig. 1.8. The 2-order ripple power $P_r$ is transferred into the decoupling circuit and the constant output power $P_o$ is maintained. Specifically, when the output power $P_o$ is smaller than the power provided by the grid $p_{ac}$, the surplus power will be absorbed by the decoupling circuit. While, when the output power $P_o$ is greater than the power provided by the grid $p_{ac}$, the inadequate power will be released by the decoupling circuit. The decoupling circuit is like an energy pool and consumes no power. As well known, the energy stored in a passive component can be expressed as $0.5CV^2$ or $0.5LI^2$. In the decoupling circuit $L$ or $C$ is small (level of mH or µF) to reduce the size, and then, the relative decoupling capacitor voltage or inductor current will swing with a large fluctuation.

Figure 1.9a, b show main circuits of the CSC and the VSC with an active power decoupling circuit. For the CSC, the equivalent circuit is shown in Fig. 1.9c. According to Kirchhoff's voltage law, if $u_{ab}$ is equal to $\tilde{u}_r$, the goal of the power decoupling will come true. The corresponding equivalent circuit of the VSC is shown in Fig. 1.9d. Similarly, according to Kirchhoff's current law, if $i_{ab}$ takes the value of $\tilde{i}_r$, the power decoupling

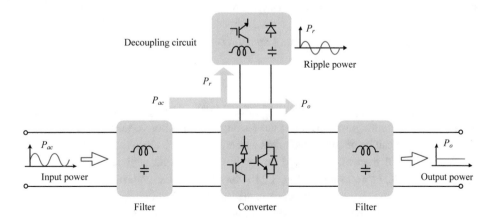

**Fig. 1.8** Schematic of active power decoupling technology

will be realized. This is to understand the power decoupling from the view of the active power filter.

## 1.4.2 Decoupling Circuit Topologies

The decoupling circuit determines the way of transferring ripple power, which is an active field of the power decoupling technique. Figure 1.10 shows a broad categorization of the decoupling circuit topologies. According to the operating characteristics of the switches and energy storage elements, they are divided into independent, dependent, and other decoupling topologies. In the independent ones, the decoupling circuit and the original circuit operate independently; while in the dependent ones, the decoupling circuit and the original circuit share the switches or energy storage elements partially or fully. The dependent decoupling topologies are achieved by horizontal or vertical multiplexing or differential connecting the dependent topologies with the original converters. The independent and dependent decoupling topologies will be presented in detail in Chap. 2. From the perspective of energy storage components, the independent decoupling circuit topologies can be further divided into capacitive storage and inductive storage ones. Besides, other decoupling topologies contain line frequency converters, matrix converters, single-phase to three-phase converters and so on.

## 1.4.3 Control Strategies

This section aims at providing an overview of prior-art and state-of-the-art decoupling control methods in active power decoupling. According to the control ideas, the decoupling control methods are classified into four kinds: power balance-based control (PBBC), harmonic suppression-based control (HSBC), volt-second balance-based/charge balance-based control (VBBC/CBBC), and virtual impedance-based control (VIBC). For each control idea, there are also a number of specific control strategies.

(1) *PBBC Method*

Suppose the 2-order ripple power is fully buffered by the decoupling capacitor or inductor. Then, according to the power balance, the decoupling capacitor voltage or the decoupling inductor current can be obtained and taken as a control target. Once the referenced voltage or current is well tracked, the ripple power is thought to be well handled. Taking the single-phase VSC as an example, Fig. 1.11 shows the specific control scheme.

The precise voltage/current reference is critical to the decoupling performance. It can be obtained via open-loop or close-loop calculation. The open-loop calculation process is briefly introduced as follows.

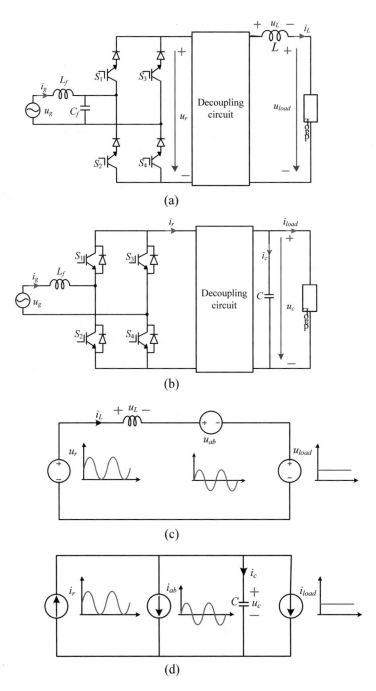

**Fig. 1.9** Current-source and voltage-source converters configured with a decoupling circuit. **a** CSC. **b** VSC. **c** Equivalent circuit of CSC. **d** Equivalent circuit of VSC

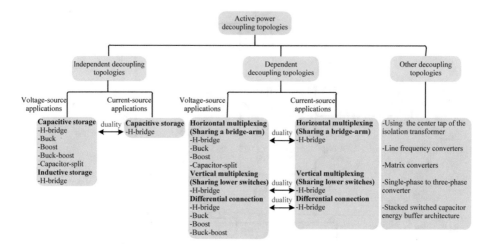

**Fig. 1.10** Classification of the decoupling topologies

**Fig. 1.11** Control scheme for
PBBC method

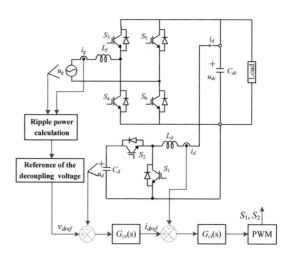

Considering $P_r$ is absorbed by the decoupling capacitor $C_d$, we can obtain the
following equation:

$$\frac{1}{2}C_d\frac{dv_d^2}{dt}=VI\cos(2\omega t+\varphi)/2 \tag{1.13}$$

By solving (1.13), the decoupling capacitor voltage $v_d(t)$ meets

$$v_d^2(t) = \frac{VI\sin(2\omega t+\varphi)}{2\omega C_d} + K \tag{1.14}$$

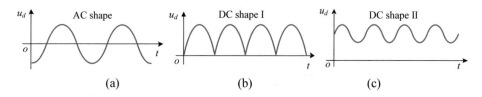

**Fig. 1.12** Decoupling capacitor voltage. **a** AC shape. **b** DC shape I. **c** DC shape II

where $K$ is a constant which affects the voltage waveforms. After some elementary calculations, $v_d(t)$ can be expressed as

$$v_d(t) = \begin{cases} V_d \sin(\omega t + \theta), & K = \frac{VI}{2\omega C_d}, \text{ AC shape} \\ V_d |\sin(\omega t + \theta)|, & K = \frac{VI}{2\omega C_d}, \text{ DC shape I} \\ \sqrt{A \sin(2\omega t + \varphi) + K}, & K = \frac{VI}{2\omega C_d}, \text{ DC shape II} \end{cases} \qquad (1.15)$$

where

$$V_d = \sqrt{\frac{VI}{\omega C_d}}, \theta = \frac{\varphi}{2} + \frac{\pi}{4}, \text{ and } A = \frac{VI}{2\omega C_d}. \qquad (1.16)$$

From (1.15), the decoupling capacitor voltage can be controlled to be AC shape or DC shape (as depicted in Fig. 1.12). When the inductor is used as the storage element, its current waveform can be analyzed similarly.

*(2) HSBC Method*

In VSC, suppose the DC-link voltage $u_{dc}$ is constant. Based on the previous analysis, the rectified output current of the ac side is expressed as (1.5). To maintain constant $u_{dc}$, there are two ways. One is to control the decoupling circuit to generate a counteraction current $i_c$ (be the same amplitude but out of phase with the ripple current $\tilde{i}_r$) to offset $\tilde{i}_r$, as shown in Fig. 1.13a. Therefore, the dc capacitor will be free of the low frequency harmonic current. In this way, the decoupling circuit acts as a classical active power filter (APF) and this control idea is called APF based control (APFC). The other is to control the decoupling circuit to generate a compensated voltage $u_c$ to offset the 2-order ripple voltage in $u_{dc}$, as shown in Fig. 1.13b, c. Then, the sum of $u_{dc}$ and $u_c$ is still constant. In this way, the decoupling circuit acts as a dynamic voltage restorer (DVR) and this control idea is called DVR based control (DVRC). Note that employing APFC or DVRC depends on the specific circuit structure. These two control strategies are most frequently used in the field of active power decoupling.

Many control strategies developed for APF applications can be used with minor modifications. And this control scheme is independent of the original rectifier/inverter control.

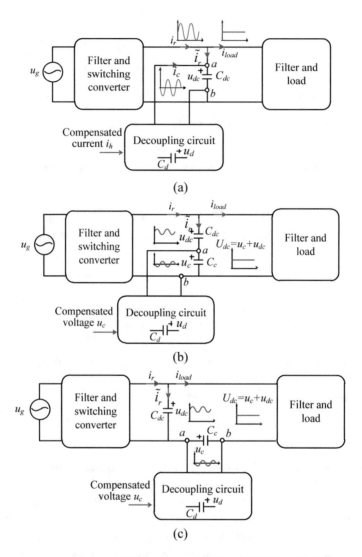

**Fig. 1.13** Control scheme for the HSBC method. **a** APFC. **b** DVRC I. **c** DVRC II

However, this control method is an intrusive approach since the rectified current $i_r$ should be sensed to extract the compensated ripple current (for the case of APFC) or a capacitor has to be inserted into the DC bus to provide a compensated voltage (for the case of DVRC). In addition, an accurate current/voltage reference during input/output power transient scenarios is almost impossible to obtain. This is because a DC component is also required in the compensated voltage/current during the transient. Besides, because of the phase delays of the filters, it takes some time for the compensated voltage/current to enter its steady-state value. When applying for rectifier, this control method is usually open loop because the decoupling effect, whether the DC bus voltage contains residual low frequency components, is not fed back. In addition, the frequency of the compensated ripple reference must be known to design the proportional-resonant (PR) or the repetitive controller.

(3) *VSBB/CBBC Method*

In terms of the load, the object of employing a decoupling circuit is to maintain a stiff bus voltage (bus current) for a VSC (CSC). From the view of the charge balance of a capacitor (volt-second balance of an inductor), the decoupling circuit keeps the charge balance of the dc bus filter capacitor (volt-second balance of the dc bus filter inductor). Based on this control concept, the decoupling capacitor voltage reference is not required and the ripple voltage or current information is not needed. Since the DC bus capacitor voltage is directly controlled by the decoupling circuit, this control is also named direct voltage control. Besides, for no dedicated ripple power decoupling controller is needed, this control is also called automatic-power-decoupling control.

To accomplish control aims, two voltage control loops are needed. Taking the buck decoupling circuit in Fig. 1.14 as an example, one voltage control is to regulate the DC bus capacitor voltage tightly. This control loop needs a high control bandwidth to achieve a fast regulation. The other is to maintain the decoupling capacitor voltage at a given voltage level $U_{dref}$. This control loop is designed to own a low bandwidth to avoid distorting the grid current. It can be found that this control concept should coordinate the decoupling control and the rectification control.

Different from PBBC and HSBC control ideas, in which the decoupling circuits only handle the ripple voltage/current of the DC bus, VBBC/CBBC control is to directly regulate the DC bus current or voltage. It is a two-stage cascaded control structure, in which the rectifier regulates the decoupling circuit capacitor voltage (regulating its average value) and the decoupling circuit maintains a stiff bus voltage/current (regulating its instantaneous value). This control concept is a close loop decoupling control and achieves excellent decoupling performance. In addition, this control has strong robustness since no precise reference is required. And the transient performance is also superior because the DC output voltage is tightly regulated by the decoupling circuit and the controller can be designed with fast response. Besides, this control concept is general and can be applied for various decoupling circuits with/without minor modification. However, the decoupling capacitor voltage needs to be unipolar with a certain dc component, storing a large portion of redundant energy. Therefore, the voltage utilization ratio is relatively low. What's worse, unstable and non-minimum phase circumstances will occur when the ripple power sweeps between the DC bus and the decoupling circuit. That increases the difficulty of controller design. Besides, since the decoupling capacitor voltage fluctuates with a large range, the DC voltage regulation system is a time-varying linear system and the stability

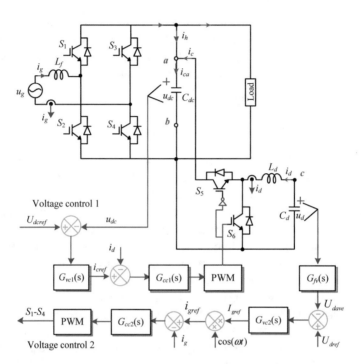

**Fig. 1.14** Applying the CBBC method for the buck decoupling circuit

is hard to establish (linear matrix inequality (LMI) or structured perturbation theory may be involved).

(4) *VIBC Method*

As introduced in Sect. 1.3, a large capacitor/inductor or an LC resonator tuned at $2f_{ac}$ resonant frequency can be used to buffer the 2-order ripple power. As well known, the output impedance of a two-terminal power electronic converter is flexible. Then, the decoupling circuit can be controlled to perform the behavior of a large capacitor/inductor or an LC resonator. From this aspect, some control schemes have been proposed.

The impedance emulation is realized by mimicking the exact voltage-current characteristic of a physical passive component. For the voltage source type converter, a current-controlled current source (CCCS) method is demonstrated in Fig. 1.15a. The low frequency current in the DC bus filter capacitor is detected and magnified $k$-fold by $G(s)$. The result is taken as the track reference of the decoupling circuit. Then, the decoupling circuit is equal to a capacitor with the value of $kC_{dc}$. The extra term $i_{virtual}$ is to compensate system losses. For the current source type converter, a voltage-controlled voltage source (VCVS) method is demonstrated in Fig. 1.15b. The low frequency voltage in the

DC bus filter inductor is detected and magnified $k$-fold by $G(s)$. Similarly, the result will be the track reference of the decoupling circuit. Then, the decoupling circuit is equal to an inductor with the value of $kL_{dc}$. In this method, the discontinuous current or voltage is sampled to avoid using a differentiator.

The details of emulating the behavior of an LC resonator with the decoupling circuit will be given in Chap. 6.

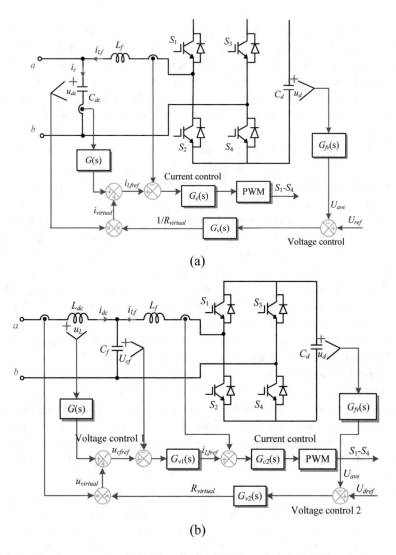

(a)

(b)

**Fig. 1.15** Application examples of VIBC method. **a** CCCS method. **b** VCVS method

VIBC is an indirect decoupling method and the decoupling effects are determined by the emulation accuracy. The decoupling circuit can be linked to the host system through the DC-link directly and no intrusive sensors and central controllers are required. Consequently, this control idea is easy to achieve plug-and-play operation.

## 1.5    Organization of the Book

The central purpose of this book is to give a comprehensive and in-depth introduction into active power decoupling technology and its most recent topics. With this book we intend to help researchers, engineers, and designers to properly select decoupling circuit topologies and control strategies according to the specific application. It should be noted that this book focuses on single-phase current source converter, and the introduced topology construction principles and the control concepts can also be extended to single-phase voltage source converters.

Chapter 2 focuses on the basic decoupling cells and the topology construction principles. The basic decoupling cells are deduced from the basic single-phase voltage-source and current-source converters. According to the type of the ripple power storage elements, they are further divided into the capacitive energy storage cell and the inductive energy storage cell. The development laws for the active power decoupling topologies are revealed from the view of "duality principle", "switches sharing", and "differential connection". This chapter is the base for the Chaps. 3, 4, 5 and 6.

Chapter 3 presents the independent decoupling circuit topology. A current-source-oriented basic decoupling cell called the asymmetric H bridge is applied to the single-phase current source rectifier. It could be viewed as a controlled voltage source in series with the DC inductor and it works with the original rectifier independently. The fundamental principle, modulation strategy, control method (PBBC control), selection of the buffer capacitor, and experimental results are introduced.

Chapter 4 is to introduce the dependent decoupling circuit topologies. They are also called switching multiplexing decoupling circuits. The introduced circuits include double capacitors and single capacitor decoupling circuits. Both of them require no additional switching devices, which are usually required in independent decoupling circuit. The topological structure, working principle, switching state, modulation strategy, control method (VBBC control), application scenarios of the two proposed decoupling circuits are investigated.

Chapter 5 aims to introduce the decoupling circuits with integrating the voltage boost function. As research continues, more and more functions have been merged into the decoupling circuits. As well known, single-phase current source rectifier is a buck-type rectifier in essence. And the DC output voltage is limited to the half peak grid voltage. This chapter presents a family of two-port switching networks to decouple the ripple power and boost the output voltage simultaneously.

Chapter 6 discusses the virtual-impedance-based decoupling control. It is realized by emulating the output voltage characteristic of the LC resonator. The specific applications in single current source rectifier and single voltage source rectifier are presented. The pros and cons of the virtual-impedance-based decoupling control are also discussed.

Chapter 7 presents the stability analysis and improvement of the single-phase current source rectifier with power decoupling circuit under constant power loads (CPLs). Single-phase AC-DC converter with the decoupling circuit is a non-linear time-periodic (NLTP) system. The classical root locus and Routh stability criterion, which are widely used in linear system, fail to work. Therefore, the harmonic state-space (HSS) modeling method is adopted to analyze the system stability. And a stability improved method of adding a modulation term to the control output reference of the decoupling circuit is introduced.

Chapter 8 presents a discrete harmonic state space (DHSS) modeling method to analyze the stability of a single-phase current source power decoupling converter with digital controller. The DHSS model can be regarded as the duality method of HSS model, remaining the autonomous nature of HSS model. It is more suitable to accurately describe the digital control system. Thus, it is an excellent candidate for analyzing the stability of NLTP systems.

## Bibliography

1. Singh, B., Singh, B. N., Chandra, A., Al-Haddad, K., Pandey, A., & Kothari, D. P. (2003). A review of single-phase improved power quality AC-DC converters. *IEEE Transactions on Industrial Electronics, 50*(5), 962–981.
2. Lacressonni, F., Cassoret, B., & Brudny, J. F. (2005). Influence of a charging current with a sinusoidal perturbation on the performance of a lead–acid battery. *Proceedings of the Institution of Electrical Engineers—Electric Power Applications, 152*(5), 1365–1370.
3. Fontes, G., Turpin, C., Astier, S., & Meynard, T. A. (2007). Interactions between fuel cells and power converters: Influence of current harmonics on a fuel cell stack. *IEEE Transactions on Power Electronics, 22*(2), 670–678.
4. Hu, H., Harb, S., Kutkut, N., Batarseh, L., & Shen, Z. J. (2013). A review of power decoupling techniques for microinverters with three different decoupling capacitor locations in PV systems. *IEEE Transactions on Power Electronics, 28*(6), 2711–2726.
5. Kim, H., & Shin, K. (2012). DESA: Dependable, efficient, scalable architecture for management of large-scale batteries. *IEEE Transactions on Industrial Informatics, 8*(2), 406–417.
6. Wang, H., & Blaabjerg, F. (2014). Reliability of capacitors for DC-link applications in power electronic converters—an overview. *IEEE Transactions on Industry Applications, 50*(5), 3569–3578.
7. Mellincovsky, M., Yuhimenko, V., Zhong, Q.-C., Peretz, M. M., & Kuperman, A. (2018). Active DC-link capacitance reduction in grid-connected power conversion systems by direct voltage regulation. *IEEE Access, 6*, 18163–18173.
8. Krein, P. T., & Balog, R. S. (2009). Cost-effective hundred-year life for single-phase inverters and rectifiers in solar and LED lighting applications based on minimum capacitance requirements and a ripple power port. In *Proceedings of the IEEE Applied Power Electronics Conference and Exposition (APEC)* (pp. 620–625).

9.  Oruganti, R., & Palaniapan, M. (2000). Inductor voltage control of buck-type single-phase AC-DC converter. *IIEEE Transactions on Power Electronics, 15*(2), 411–416.

10. Gemmen, R. S. (2003). Analysis for the effect of inverter ripple current on fuel cell operating condition. *Journal of Fluids Engineering, 125*(3), 576–585.

11. Wang, R., Wang, F., Boroyevich, D., Burgos, R., Lai, R., Ning, P., & Rajashekara, K. (2003). A high power density single-phase PWM rectifier with active ripple energy storage. *IEEE Transactions on Power Electronics, 26*(5), 1430–1443.

12. Dusmez, S., & Khaligh, A. (2014). Generalized technique of compensating low-frequency component of load current with a parallel bidirectional DC-DC converter. *IEEE Transactions on Power Electronics, 29*(11), 5892–5904.

13. Li, S., Tan, S. C., Lee, C., Waffenschmidt, E., Hui, S. Y. R., & Tse, C. K. (2015). A survey, classification and critical review of light-emitting diode drivers. *IEEE Transactions on Power Electronics,* (99), 1.

14. Li, S., Qi, W., Tan, S., Hui, S. Y., & Tse, C. K. (2018). A general approach to programmable and reconfigurable emulation of power impedances. *IEEE Transactions on Power Electronics, 33*(1), 259–271.

15. Chen, W., & Hui, S. Y. R. (2012). Elimination of an electrolytic capacitor in AC-DC light-emitting diode (LED) driver with high input power factor and constant output current. *IEEE Transactions on Power Electronics, 27*(3), 1598–1607.

16. Huai, W., Liserre, M., & Blaabjerg, F. (2003). Toward reliable power electronics: Challenges, design tools, and opportunities. *IEEE Transactions on Power Electronics, 7*(2), 17–26.

17. Ohnuma, Y., & Itoh, J. I. (2014). A novel single-phase buck PFC AC–DC converter with power decoupling capability using an active buffer. *IEEE Transactions on Industry Applications, 50*(3), 1905–1914.

18. Sun, Y., Liu, Y., Su, M., Xiong, W., & Yang, J. (2016). Review of active power decoupling topologies in single-phase systems. *IEEE Transactions on Power Electronics, 31*(7), 4778–4794.

19. Liu, Y., Zhang, W., Sun, Y., Su, M., Xu, G., & Dan, H. (2021). Review and comparison of control strategies in active power decoupling. *IEEE Transactions on Power Electronics, 36*(12), 14436–14455.

# Basic Decoupling Cells and Topology Construction Principles

**2**

Decoupling circuit topologies are the hardware foundation to deal with the ripple power. Some basic performance indices, such as cost, voltage stress, power losses, and control complexity, are directly related to the decoupling circuit. This chapter detailly introduces the basic decoupling cells and topology construction principles. The basic decoupling cells includes the voltage-source-oriented and the current-source-oriented basic decoupling cells. The development laws for the active power decoupling topologies are revealed from the view of "duality principle", "switches sharing", and "differential connection". In this chapter, the basic decoupling cells and the development laws are introduced in Sects. 2.1 and 2.2. Section 2.3 concludes this chapter.

## 2.1 Basic Decoupling Cells

This section is to introduce the basic decoupling cells and pave the way for the following sections. Basic decoupling cells are the modules which are able to buffer the ripple power. They are connected to the original converters in series, parallel, or other ways. According to the scope of applications, they can mainly be categorized into two groups: the voltage-source oriented basic cells and the current-source oriented basic cells. And both of them can be further divided into two types: capacitive energy storage and inductive energy storage.

© The Author(s), under exclusive license to Springer Nature Switzerland AG 2023  21
Y. Liu, *Active Power Decoupling Technology in Single-Phase Current-Source Converters*,
Synthesis Lectures on Power Electronics, https://doi.org/10.1007/978-3-031-21270-3_2

### 2.1.1  Voltage-Source-Oriented Basic Decoupling Cells

(1) *Capacitive Energy Storage*

For the capacitive energy storage, the ripple energy is stored in the electrostatic field of capacitors. Figure 2.1 shows two kinds of voltage-source oriented basic decoupling cells. Both of them have the same H-bridge structure but different features. The decoupling capacitor $C_b$ in Fig. 2.1a could withstand alternative voltages. However, the absolute value of the capacitor voltage must be lower than the DC terminal voltage $u_{ab}$. Usually, to buffer the twice ripple power the waveforms of the capacitor voltage and current are controlled to be the shape as illustrated in Fig. 2.2a. The decoupling capacitor in Fig. 2.1b could only withstand a DC voltage which must be higher than the corresponding terminal voltage $u_{ab}$. Under normal condition, its corresponding waveforms are shown in Fig. 2.2b. Therefore, considering the same operation condition, the basic decoupling cell in Fig. 2.1a has low voltage stress but needs a capacitor with relatively large capacitance; while the basic decoupling cell in Fig. 2.1b has to suffer high voltage stress but only needs a capacitor with relatively small capacitance.

In fact, if the capacitor of the basic cell in Fig. 2.1a is a polarized capacitor, the basic cell can be reduced to the buck-type decoupling cell as shown in Fig. 2.3a. Due to the voltage polarity constraint, the possible waveforms of the capacitor in the buck-type basic decoupling cell are shown in Fig. 2.2c, d. It is clear that the voltage across the energy storage capacitor in Fig. 2.2c is not as smooth as that in Fig. 2.2d. From the point view of control, the waveforms in Fig. 2.2c are more difficult to realize than those in Fig. 2.2d. However, the capacity of the energy storage capacitor can be utilized fully in Fig. 2.2c.

Similarly, if the terminal voltage $u_{ab}$ of the basic cell in Fig. 2.1b is a DC voltage, the basic cell can be reduced to the boost-type basic decoupling cell as illustrated in Fig. 2.3b. Such a simplification has no impact on the value of capacitance. In addition, that reduces cost and the system complexity. According to the characteristics of the basic cells in Fig. 2.3a, b, it can be concluded that almost all the bidirectional DC-DC circuits are the candidates for the basic decoupling cells, for example, the bidirectional buck-boost circuit shown in Fig. 2.3c. A merit of buck-boost-type basic decoupling cell is that its capacitor

**Fig. 2.1** H-bridge basic decoupling cell with a capacitor as the energy storage unit. **a** With alternative capacitor voltage and DC terminal voltage. **b** With alternative terminal voltage and DC capacitor voltage

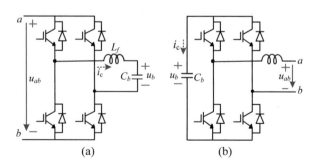

(a)                              (b)

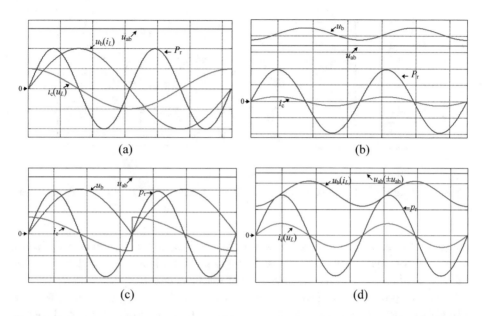

**Fig. 2.2** Waveforms of the decoupling capacitor voltage $u_b$ (decoupling inductor current $i_L$), decoupling capacitor current $i_c$ (decoupling inductor voltage $u_L$), ripple power $P_r$ and the terminal voltage $u_{ab}$ when the capacitor voltage (decoupling inductor current $i_L$) is controlled under different references. **a** $u_b$ ($i_L$) is controlled to be sine wave. **b** $u_b$ is controlled to be higher than the terminal voltage. **c** $u_b$ is controlled to be full-wave rectified sine wave. **d** $u_b$ is controlled to be lower than the terminal voltage ($i_L$ is controlled to be a sine waveform with a predetermined DC bias)

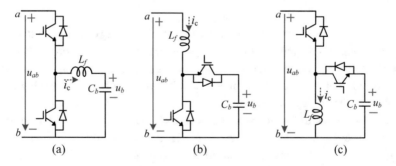

**Fig. 2.3** Basic decoupling cell configured with bidirectional DC-DC circuits. **a** Buck-type basic decoupling cell. **b** Boost-type basic decoupling cell. **c** Buck-boost-type basic decoupling cell

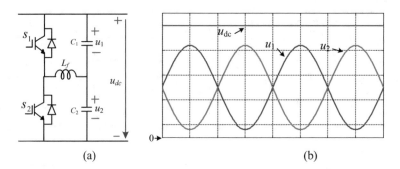

(a)                                                          (b)

**Fig. 2.4**  Capacitor-split basic decoupling cell. **a** Schematic diagram. **b** Waveforms of the DC capacitors

voltage has no limitations and can be lower or higher than the terminal voltage $u_{ab}$, which widens the potential applications of the decoupling circuit. However, the drawbacks are the reduced efficiency and increased volume of filter inductor $L_f$.

The ripple power, in all of the aforementioned basic decoupling cells, is buffered by a single capacitor. Figure 2.4a shows a capacitor-split basic decoupling cell, in which the ripple power is buffered by two capacitors. Capacitors $C_1$ and $C_2$ are identical and connected in series. Usually, the capacitor voltage waveforms are regulated as shown in Fig. 2.4b by controlling switches $S_1$ and $S_2$. The capacitors $C_1$ and $C_2$ play dual roles of mitigating the ripple power and filtering switching ripple.

(2) *Inductive Energy Storage*

As well known, inductors can also be used as a kind of energy storage unit for electric energy. And superconducting magnetic energy storage is a classic example. For the basic decoupling cell based on the inductive energy storage, the ripple energy is stored in the electromagnetic field of inductors. Figure 2.5a, b show two basic decoupling cells with an inductor as the energy storage unit. The inductor in Fig. 2.5a is an AC inductor, and its current can be controlled to be a sine waveform without any DC bias, as shown in Fig. 2.2a. However, the inductor current $i_L$ in Fig. 2.5b is usually controlled to be a sine waveform with a predetermined DC bias to keep the unipolarity, as shown in Fig. 2.2d. In Fig. 2.5b $u_{ab}$ can be bipolar. However, in many occasions, it only needs to be connected to a DC voltage, and then its simplified version is obtained as shown in Fig. 2.5c.

## 2.1.2  Current-Source-Oriented Basic Decoupling Cells

Regarding to basic decoupling cells applied in CSCs, usually a capacitor is selected as the ripple energy storage element, which guarantees the minimal order of the decoupling

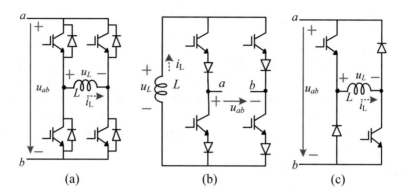

**Fig. 2.5** H-bridge basic decoupling cell with an inductor as the energy storage unit. **a** With AC inductor current and DC terminal voltage. **b** With AC terminal voltage and DC inductor current. **c** With DC inductor current and terminal voltage

cells. According to the duality principle, two basic decoupling cells suited for CSCs are derived from the basic cells in Fig. 2.5a, b, which are illustrated in Fig. 2.6a, b.

The capacitor in Fig. 2.6a should withstand AC voltage but the terminal current $i_{dc}$ is limited to be positive. The contrary is the case in Fig. 2.6b. If the capacitor in Fig. 2.6a only needs to withstand a DC voltage or the terminal current $i_{dc}$ of the cell in Fig. 2.6b is unidirectional, both of them can be simplified into the basic cell shown in Fig. 2.6c.

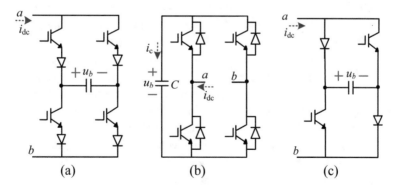

**Fig. 2.6** Current-source oriented basic decoupling cells. **a** With AC capacitor voltage and unidirectional terminal current. **b** With DC capacitor voltage and AC terminal current. **c** With DC capacitor voltage and terminal current

## 2.2    Topology Construction Principles

Usually, the decoupling circuit topologies consist of basic decoupling cells and the original converters. In some of them the basic decoupling cells and the original converters work independently. In some other cases, the basic decupling cell shares switches with the original converter partially and even completely. As a result, the shared switches work coordinately to achieve the goal of power decoupling as well as power conversion. Besides, there are some decoupling circuit topologies formed by the differential concept, in which the decoupling capacitors play twofold roles of buffering the ripple power and filtering the switching harmonics.

### 2.2.1    Independent Decoupling Topologies

The independent decoupling topologies mean that the single-phase converters and the basic decoupling cells are able to operate independently. Usually, the basic decoupling cell is connected to the DC-link in parallel or series. The single-phase converter is responsible for regulating the DC-link voltage/current and basic decoupling cell is to deal with the ripple power. The introduction of the additional basic decoupling cell will not change the operation point of the original single-phase converter. Meanwhile, the control methods and modulation strategies for the original single-phase converter and the basic decoupling cell can be designed independently. Some independent decoupling circuit topologies will be reviewed in the following.

Figure 2.7 shows a framework of the general power conditioner with H-bridge basic decoupling cell in Fig. 2.6b in parallel and series. The basic idea of the parallel decoupling concept is to inject compensation current to the coupling point, which is able to prevent the twice current ripple from flowing into DC capacitor. That idea is equivalent to the principle of the parallel active power filter. The basic idea of the series decoupling concept is to inject compensation voltage in series to mitigate the pulsed voltage in DC-link voltage $u_c$ caused by the twice ripple power, which is equivalent to the principle of the series active power filter. Although the DC-link voltage $u_d$ fed to right-side converter in Fig. 2.7b is smooth, the low-frequency ripple voltage in DC-link voltage $u_c$ still exists. As a result, the effect of low-frequency ripple voltages on the MPPT efficiency of PV modules still needs to be taken into account.

Figure 2.8 shows an independent decoupling topology tailored for current-source converters. It is developed by connecting the H-bridge basic decoupling cell in Fig. 2.6c in series with the DC-link. It acts as a controlled voltage source to compensate the ripple component in the rectified output voltage, which is a classic example for the series decoupling concept. And it will be detailly introduced in the next Chapter.

**Fig. 2.7** Framework of general power conditioner with H-bridge basic decoupling cell. **a** In parallel. **b** In series

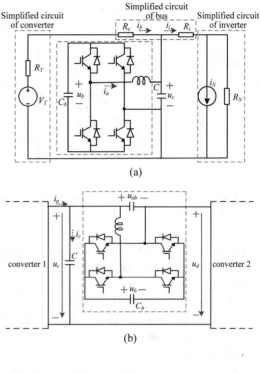

**Fig. 2.8** Single-phase current-source converter with H-bridge basic decoupling cell

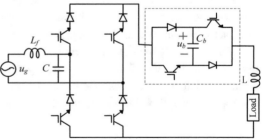

## 2.2.2 Dependent Decoupling Topologies

This section will review some dependent decoupling topologies in which the basic decoupling cell shares power semiconductor devices with the original converter partially or even fully. In the following section, the dependent decoupling topologies are also called "switch-multiplexing decoupling topologies".

Figure 2.9 shows switch-multiplexing decoupling topology resulting from the topology in Fig. 2.5a. There are two different switch-multiplexing decoupling topologies. In Fig. 2.9a, the middle bridge-arm is shared by the original converter and the decoupling cell. The shared bridge-arm undertakes two tasks simultaneously: rectification/inversion

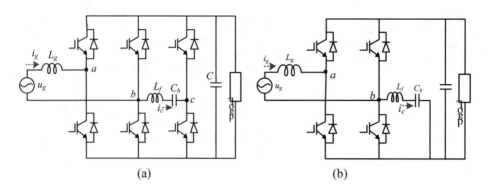

**Fig. 2.9** Switch-multiplexing decoupling topology resulting from the topology in Fig. 2.7a. **a** Version I. **b** Version II

and ripple power buffering. One obvious advantage is that two active switches have been saved. If the rectifier bridge and the H-bridge basic decoupling cell are fully merged, the decoupling circuit showed in Fig. 2.9b can be obtained. Consequently, no extra switch is added, which is expected in practice. However, both of them have some drawbacks, for example, the viable operating range of the DC-link voltage becomes smaller and the voltage stress may become higher than those in the dependent decoupling topology.

Figure 2.10 shows switch-multiplexing decoupling topology resulting from the topology in Fig. 2.8. There are also two different switch-multiplexing decoupling topologies. The switches are both used to achieve rectification and ripple power buffering. And no extra active switch is added. In Fig. 2.10a, there are two decoupling capacitors $C_1$ and $C_2$, which work alternately. After removing one capacitor, the switch-multiplexing decoupling topology in Fig. 2.10b can be deduced. Then, only one decoupling capacitor is needed, which saves the cost. However, the operation constraint is strengthened. These two topologies will be introduced in detail in Chap. 4.

### 2.2.3  Differential Connection

The aforementioned switch-multiplexing decoupling circuits are summarized from the point of the switching sharing concept. Besides, the differential connection is also an important way to construct decoupling circuit topologies. As well known, many classical AC-DC converters are derived from basic DC-DC converters by differential connection. Actually, AC-DC converters formed by differential connection have the inherent capability of ripple power decoupling via controlling the common mode voltages of the output filter capacitors.

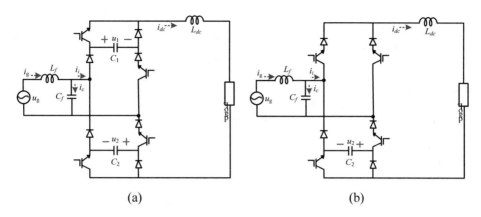

**Fig. 2.10** Switch-multiplexing circuit topology with double decoupling capacitors (**a**) and single decoupling capacitor (**b**)

Figure 2.11a shows a possible decoupling solution formed by two identical full bridge inverters in differential connection way. In steady state the capacitor voltages $u_1$ and $u_2$ satisfy

$$\begin{cases} C_1 u_1 \dfrac{du_1}{dt} + C_2 u_2 \dfrac{du_2}{dt} = P_r \\ u_1 - u_2 = u_g \end{cases} \tag{2.1}$$

It can be found that the differential common voltage is the gird voltage and the ripple power $P_r$ is also buffered by the filter capacitors. Actually, in Fig. 2.11a the first or the fourth bridge-arm (from left to right) is redundant. After removing the redundant one, a simplified version shown in Fig. 2.11b is obtained. Through the analysis of circuit in Fig. 2.11b, if the output terminal of the second bridge-arm is clamped to the negative bus-bar, the converter as shown in Fig. 2.11c is obtained. The main difference between converters in Fig. 2.11b, c is that the voltages across the $C_1$ and $C_2$ can be controlled to be AC voltages in the former, while only DC voltages in the latter. With regard to decoupling topology in Fig. 2.11c, actually, it is composed of two buck DC-DC circuits, which can be called as the buck-type differential inverters. Obviously, there exist other cases such as the boost-type concept shown in Fig. 2.11d and buck-boost-type concept. Both of them could also achieve ripple power decoupling with proper control methods.

## 2.3 Conclusion

This chapter introduces basic decoupling cells and decoupling topology construction principles. Active power decoupling topologies with the features of low cost, low volume

**Fig. 2.11** Decoupling
topologies with differential
connection concept. **a** Two
H-bridge basic decoupling
cells in Fig. 2.1a connected in
differential way.
**b** Switch-multiplexing version
resulting from Fig. 2.11a.
**c** Buck-type differential
inverter. **d** Boost-type
differential inverter

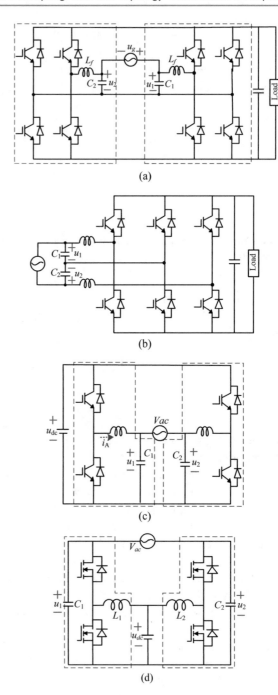

and weight, high efficiency, high reliability, and high performance are always expected. However, there is no such thing as a free lunch. For independent decoupling topologies, decoupling circuits and original circuits are not interfered with each other, which leads to flexible designs in control methods and modulation strategies. However, this kind of decoupling method usually involves a lot of additional power semiconductor devices, which increases cost significantly. On the other hand, the power decoupling topologies with switch-multiplexing benefit from fewer power semiconductor devices. However, they usually suffer from more constraints, such as increased voltage/current stress, reduced operation range, increased volume, and complicated control algorithms. And each active power decoupling topology has its advantages and disadvantages, so an evaluation method under multi-objective optimization framework should be studied to help to construct or select a proper topology for a specific application.

## Bibliography

1. Tian, B., Harb, S., & Balog, R. S. (2014). Ripple-port integrated PFC rectifier with fast dynamic response. In *Proceedings of the IEEE MWSCAS*, College Station, TX (pp. 781–784).
2. Lee, C. T., Chen, Y. M., Chen, L. C., & Cheng, P. T. (2012). Efficiency improvement of a DC-AC converter with the power decoupling capability. In *Proceedings of the IEEE APEC*, Orlando, FL (pp. 1462–1468).
3. Zonxiang, C., Chao, L., Fenghua, Y., LiuSheng, G. (2012). A single-phase grid-connected inverter with an active power decoupling circuit. In *Proceedings of the IEEE CCDC*, Taiyuan (pp. 2806–2810).
4. Zhu, G. R., Tan, S. C., Chen, Y., & Tse, C. K. (2013). Mitigation of low-frequency current ripple in fuel-cell inverter systems through waveform control. *IEEE Transactions on Power Electronics, 28*(2), 779–792.
5. Han, H., Liu, Y., Sun, Y., Su, M., & Xiong, W. (2015). Single-phase current source converter with power decoupling capability using a series-connected active buffer. *IET Power Electronics, 8*(5), 700–707.
6. Li, H., Zhang, K., & Zhao, H. (2012). DC-link active power filter for high-power single-phase PWM converters. *Journal of Power Electronics, 12*(3), 458–467.
7. Li, H., Zhang, K., Zhao, H., Fan, S., & Xiong, J. (2013). Active power decoupling for high-power single-phase PWM rectifiers. *IEEE Transactions on Power Electronics, 28*(3), 1308–1319.
8. Wang, H., Chung, H. S. H., & Liu, W. (2014). Use of a series voltage compensator for reduction of the DC-link capacitance in a capacitor-supported system. *IEEE Transactions on Power Electronics, 29*(3), 1163–1175.
9. Serban, I. (2015). Power decoupling method for single-phase h-bridge inverters with no additional power electronics. *The IEEE Transactions on Industrial Electronics, (99)*, 1.
10. Vitorino, M. A., & de Rossiter Correa, M. B. (2014). Compensation of DC-link oscillation in single-phase VSI and CSI converters for photovoltaic grid connection. *IEEE Transactions on Industry Applications, 50*(3), 2021–2028.
11. Jang, M., & Agelidis, V. G. (2011). A minimum power-processing-stage fuel-cell energy system based on a boost-inverter with a bidirectional backup battery storage. *IEEE Transactions on Power Electronics, 26*(5), 1568–1577.

12. Jang, M., Ciobotaru, M., & Agelidis, V. G. (2012). A single-stage fuel cell energy system based on a buck–boost inverter with a backup energy storage unit. *IEEE Transactions on Power Electronics, 27*(6), 2825–2834.
13. Su, M., Pan, P., Long, X., Sun, Y., & Yang, J. (2014). An active power-decoupling method for single-phase AC-DC converters. *IEEE Transactions on Industrial Informatics, 10*(1), 461–468.
14. Krein, P. T., Balog, R. S., & Mirjafari, M. (2012). Minimum energy and capacitance requirements for single-phase inverters and rectifiers using a ripple port. *IEEE Transactions on Power Electronics, 27*(11), 4690–4698.
15. Chen, R., Liu, Y., & Peng, F. Z. (2015). DC capacitor-less inverter for single-phase power conversion with minimum voltage and current stress. *IEEE Transactions on Power Electronics, 30*(10), 5499–5507.
16. Wai, R. J., & Lin, C. Y. (2010). Active low-frequency ripple control for clean-energy power-conditioning mechanism. *IEEE Transactions on Industrial Electronics, 57*(11), 3780–3792.
17. Wang, R., Wang, F., Boroyevich, D., Burgos, R., Lai, R., Ning, P., & Rajashekara, K. (2011). A high power density single-phase PWM rectifier with active ripple energy storage. *IEEE Transactions on Power Electronics, 26*(5), 1430–1443.
18. Dusmez, S., & Khaligh, A. (2014). Generalized technique of compensating low-frequency component of load current with a parallel bidirectional DC-DC converter. *IEEE Transactions on Power Electronics, 29*(11), 5892–5904.
19. Harb, S., Mirjafari, M., & Balog, R. S. (2013). Ripple-port module-integrated inverter for grid-connected PV applications. *IEEE Transactions on Industry Applications, 49*(6), 2692–2698.
20. Mazumder, S. K., Burra, R. K., & Acharya, K. (2007). A ripple-mitigating and energy-efficient fuel cell power-conditioning system. *IEEE Transactions on Power Electronics, 22*(4), 1437–1452.
21. Li, S., Zhu, G. R., Tan, S. C., & Hui, S. Y. (2015). Direct AC-DC rectifier with mitigated low-frequency ripple through inductor-current waveform control. *IEEE Transactions on Power Electronics, 30*(8), 4336–4348.
22. Wang, S., Ruan, X., Yao, K., Tan, S., Yang, Y., & Ye, Z. (2012). A flicker-free electrolytic capacitor-less AC–DC LED driver. *IEEE Transactions on Power Electronics, 27*(11), 4540–4548.
23. Larsson, T., & Ostlund, S. (1995). Active DC-link filter for two frequency electric locomotives. In *Proceedings of International Conference on Electric Railways in a United Europe*, Amsterdam (pp. 97–100).
24. Shimizu, T., Fujita, T., Kimura, G., & Hirose, J. (1997). A unity power factor PWM rectifier with DC ripple compensation. *IEEE Transactions on Industrial Electronics, 44*(4), 447–455.
25. Shimizu, T., Jin, Y., & Kimura, G. (2000). DC ripple current reduction on a single-phase PWM voltage-source rectifier. *IEEE Transactions on Industry Applications, 36*(4), 1419–1429.
26. Cai, W., Liu, B., Duan, S., & Jiang, L. (2014). An active low-frequency ripple control method based on the virtual capacitor concept for BIPV systems. *IEEE Transactions on Power Electronics, 29*(4), 1733–1745.
27. Cai, W., Jiang, L., Liu, B., & Zou, C. (2015). A power decoupling method based on four-switch three-port DC-DC-AC converter in dc microgrid. *IEEE Transactions on Industry Applications, 51*(1), 336–343.
28. Liu, W., Wang, K., Chung, H., & Chuang, S. (2015). Modeling and design of series voltage compensator for reduction of DC-link capacitance in grid-tie solar inverter. *IEEE Transactions on Power Electronics, 30*(5), 2534–2548.
29. Ming, W., Zhong, Q., & Zhang, X. (2015). Transformerless single-phase rectifiers with significantly reduced capacitance. *IEEE Transactions on Power Electronics*, in press.

30. Cao, X., Zhong, Q., & Ming, W. (2015). Ripple eliminator to smooth DC-Bus voltage and reduce the total capacitance required. *IEEE Transactions on Power Electronics, 62*(4), 2224–2235.

31. Ma, X., Wang, B., Zhao, F., Qu, G., Gao, D., & Zhou, Z. (2002). A high power low ripple high dynamic performance DC power supply based on thyristor converter and active filter. In *Proceedings of the IEEE 28th Annual Conference of the IEEE Industrial Electronics Society* (pp. 1238–1242).

32. Tang, Y., & Blaabjerg, F. (2015). A component-minimized single-phase active power decoupling circuit with reduced current stress to semiconductor switches. *IEEE Transactions on Power Electronics, 30*(6), 2905–2910.

33. Tang, Y., Zhu, D., Jin, C., Wang, P., & Blaabjerg, F. (2015). A three-level quasi-two-stage single-phase PFC converter with flexible output voltage and improved conversion efficiency. *IEEE Transactions on Power Electronics, 30*(2), 717–726.

34. Tang, Y., Blaabjerg, F., Loh, P. C., Jin, C., & Wang, P. (2015). Decoupling of fluctuating power in single-phase systems through a symmetrical half-bridge circuit. *IEEE Transactions on Power Electronics, 30*(4), 1855–1865.

35. Tang, Y., Qin, Z., Blaabjerg, F., & Loh, P. C. (2014). A dual voltage control strategy for single-phase PWM converters with power decoupling function. *IEEE Transactions on Power Electronics, (99)*, 1.

36. Yang, Y., Ruan, X., Zhang, L., He, J., & Ye, Z. (2014). Feed-forward scheme for an electrolytic capacitor-less AC-DC LED driver to reduce output current ripple. *IEEE Transactions on Power Electronics, 29*(10), 5508–5517.

37. Sun, Y., Liu, Y., Su, M., Xiong, W., & Yang, J. (2016). Review of active power decoupling topologies in single-phase systems. *IEEE Transactions on Power Electronics, 31*(7), 4778–4794.

# Independent Decoupling Topology for Current-Source Rectifier

In the early stages, independent decoupling topologies attracted much research interest. And many dependent decoupling circuit variations are deduced based on them. This chapter will introduce a classical independent decoupling circuit applied to the single-phase current source converter (SCSC). It could be viewed as a controlled voltage source in series with the DC inductor, and work with the SCSC independently. That facilitates the separate design of the modulation schemes and the control algorithms for the power decoupling circuit and the SCSC. The fundamental principle of the proposed converter is analyzed in Sect. 3.1. To guarantee high input current quality, a control method, in which the input current is treated as a virtual control input, is presented in Sect. 3.2. The voltage reference and the capacitance of the decoupling (buffer) capacitor are investigated in Sects. 3.1 and 3.2. Section 3.4 gives the simulations and experimental results. And in Sect. 3.5 some other independent decoupling topologies developed for SCSC are introduced and discussed.

## 3.1 Circuit Topology

### 3.1.1 Circuit Configuration

The topology of the SCSC with power decoupling function is shown in Fig. 3.1. It is constructed by inserting an active buffer circuit (ABC) to the DC bus of the conventional SCSC. The buffer circuit consists of two switching devices ($S_5$, $S_6$), two diodes ($D_5$, $D_6$), and a capacitor ($C_d$). As seen, the power pulsation with twice the power supply frequency is absorbed by the active buffer capacitor $C_d$. Consequently, the constant power feeds the DC load.

© The Author(s), under exclusive license to Springer Nature Switzerland AG 2023
Y. Liu, *Active Power Decoupling Technology in Single-Phase Current-Source Converters*,
Synthesis Lectures on Power Electronics, https://doi.org/10.1007/978-3-031-21270-3_3

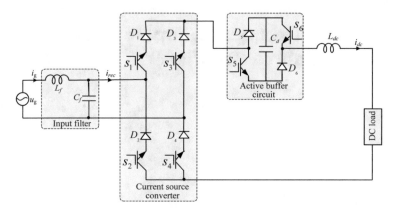

**Fig. 3.1**  Topology of the SCSC an active buffer circuit

As shown in Fig. 3.2, four operation modes exist in ABC. In mode 1, $S_5$ is turned off and $S_6$ is turned on; while in mode 3, $S_5$ is turned on and $S_6$ is turned off. Both modes are equivalent in function, and the buffer capacitor is disconnected from the main circuit and the buffer circuit works as a freewheeling path. In mode 2, both $S_5$ and $S_6$ are turned off, then the buffer capacitor is charged and the excess energy provided by the grid is absorbed. In contrast, in mode 4, both $S_5$ and $S_6$ are turned on, then the buffer capacitor is discharged and the insufficient energy required by the DC load is supplemented.

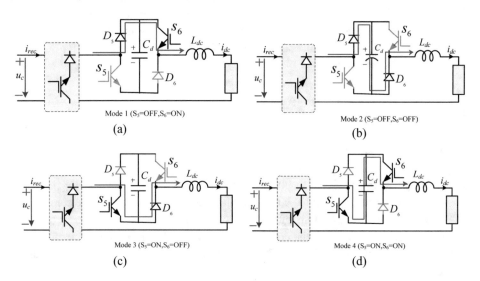

**Fig. 3.2**  Operation modes of ABC

### 3.1.2 Operation Principle

Assume that the input voltage is sinusoidal with the amplitude $V$ and angular frequency $\omega$. It is expressed as

$$u_g = V\cos(\omega t) \tag{3.1}$$

The input current $i_g$ is

$$i_g = I\cos(\omega t + \varphi) \tag{3.2}$$

where $\varphi$ is the displacement angle, and $I$ is the amplitude.

Then the converter's instantaneous input power is expressed as

$$p_{ac} = \tfrac{1}{2}[VI\cos(\varphi) + VI\cos(2\omega t + \varphi)] \tag{3.3}$$

Obviously, the second term in (3.3) is a ripple power with twice the grid frequency. In most applications, the load consumes constant power. Thus, the ripple power must be buffered to avoid the distortion in the DC current and even the input current. This task will be completed by controlling ABC properly. If ignoring the power losses caused by semiconductor devices and the input filter and assuming the ripple power is buffered by the capacitor $C_d$ in ABC, the following equation holds,

$$\frac{1}{2}C_d u_d^2(t) - \frac{1}{2}C_d u_d^2(t_0) = \int_{t_0}^{t} \frac{VI\cos(2\omega t + \varphi)}{2}dt \tag{3.4}$$

where $u_d$ is the capacitor voltage. By integrating both sides of Eq. (3.4) with respect to the time, then

$$u_d^2 = \overline{u}_d^2 + \frac{VI}{2\omega C_d}\sin(2\omega t + \varphi) \tag{3.5}$$

where $\overline{u}_d$ is the DC component.

Recalling that $u_d$ is positive, we obtain

$$u_d = \sqrt{\overline{u}_d^2 + \frac{VI\sin(2\omega t + \varphi)}{2\omega C_d}} \tag{3.6}$$

It is clear that $\overline{u}_d$ is a degree of freedom, and it satisfies the following constraint,

$$\overline{u}_d \geq \sqrt{\frac{VI}{2\omega C_d}} \tag{3.7}$$

Moreover, the low-frequency capacitor current can be expressed as

**Fig. 3.3** Operating waveforms

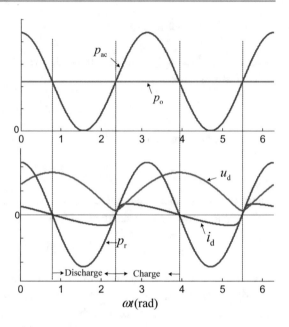

$$i_d = \frac{VI\cos(2\omega t + \varphi)/2}{\sqrt{\bar{u}_d^2 + \frac{VI\sin(2\omega t + \varphi)}{2\omega C_d}}} \tag{3.8}$$

For further showing the working principle, operating waveforms of the introduced converter are presented in Fig. 3.3. As seen, when input power $p_{ac}$ is larger than the output power $p_o$, the capacitor current $i_d$ is positive and $u_d$ increases, and the excess input power flows into the decoupling capacitor. When the AC side cannot feed the load adequate energy, the capacitor current $i_d$ is negative and $u_d$ decreases, then insufficient part is provided by the decoupling capacitor.

## 3.2    Modeling and Control

### 3.2.1    Modeling

According to Fig. 3.1, the average model of the proposed converter is formulated as follows:

$$L_f \frac{di_g}{dt} = u_g - u_c \tag{3.9}$$

$$C_f \frac{du_c}{dt} = i_g - i_{rec} \tag{3.10}$$

$$L_{dc}\frac{di_{dc}}{dt} = d_r u_c - d_d u_d - u_{dc} \tag{3.11}$$

$$C_d\frac{du_d}{dt} = i_{dc} d_d \tag{3.12}$$

$$i_{rec} = d_r i_{dc} \tag{3.13}$$

$$d_r = d_1 - d_2 \tag{3.14}$$

$$d_d = 1 - d_5 - d_6 \tag{3.15}$$

where $u_c$ is the terminal voltage of capacitor $C_f$, $u_{dc}$ is the voltage across the DC load, $i_{dc}$ is the DC current flowing through inductor $L_{dc}$, and $d_i$ is the duty ratio of switch $S_i$ ($i$ = 1, 2, …, 6). From (3.13), $d_r$ can be used to control the input current. $d_d$ can be used to control the capacitor voltage $u_d$ in ABC. Moreover, both $d_r$ and $d_d$ are defined on the interval $[-1, 1]$.

### 3.2.2   Controller Design

To complete the power decoupling, based on the above results, the capacitor voltage $u_d$ is controlled to track its reference as shown in (3.6) accurately. Here, a simple proportional control is adopted, and then according to (3.12), $d_d$ is selected as follows:

$$d_d = \frac{C_d}{i_{dc}}[\dot{u}_d^* + k(u_d^* - u_d)] \tag{3.16}$$

where $u_d^*$ is the reference of $u_d$, and $\dot{u}_d^*$ is the time derivative of $u_d^*$, which serves as a forward compensation in the control. Substituting (3.16) into (3.12), then

$$\dot{e}_d = -k e_d \tag{3.17}$$

where $e_d = u_d^* - u_d$, and $k > 0$. It is clear that $e_d$ convergences to zero asymptotically.

In the next, the control objectives are to achieve a given constant DC current and the sinusoidal grid current. The former is accomplished by controlling the ABC and the latter by controlling the SCSC.

Both sides of (3.11) are multiplied by $i_{dc}$, then the following expression is obtained,

$$\frac{L_{dc}}{2}\frac{dx}{dt} = i_{rec}u_c - P_r - P_o \tag{3.18}$$

where $x = i_{dc}^2$, $P_r = i_{dc}u_d d_d$ and $P_o = i_{dc}u_{dc}$. In order to obtain the sinusoidal grid current, the following equation should be satisfied,

$$i_{rec} = I \cos(\theta + \varphi) \tag{3.19}$$

where $\theta$ is the phase of $u_c$, which is obtained by the digital phase lock loop (PLL). $I$ is the control input, which will be designed latter.

If ignoring the effect of the input filter, $u_c$ could be approximated to $V \cos(\theta)$. By substituting (3.19) into (3.18), one has

$$\frac{L_{dc}}{2} \frac{dx}{dt} = \frac{1}{2} I V [\cos(\varphi) + \cos(2\theta + \varphi)] - P_r - P_o \tag{3.20}$$

The right side of (3.20) is a periodic function. To facilitate the design of the controller, the periodic averaging method is used here. The average differential equation is as following

$$L_{dc} \frac{d\overline{x}}{dt} = I V \cos(\varphi) - 2P_o \tag{3.21}$$

where $\overline{x}$ is obtained by a moving average filter (MAF) in implementation. The Eq. (3.21) is a linear first order differential equation, and the control law for $I$, is designed as

$$I(s) = (k_p + \frac{k_i}{s})(x^*(s) - \overline{x}(s)) \tag{3.22}$$

The overall control block diagram is shown in Fig. 3.4. Note that the DC current $i_{dc}$ as shown in Fig. 3.4 appears in denominator. Therefore, it results in singularity during starting up, which could be avoided by replacing it with its reference value during startup.

**Fig. 3.4** Block diagram of the control scheme

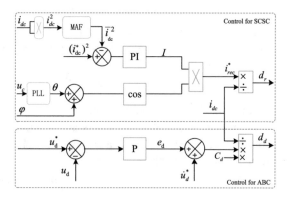

### 3.2.3  Modulation Strategy

To ensure the minimum switching loss with a fixed switching frequency, the combination of switches is restrained further as follows: if $d_r > 0$, $d_1 = 1$; otherwise, $d_1 = 0$. If $d_d > 0$, $d_6 = 0$; otherwise, $d_6 = 1$. The duty ratio of each switch is summarized in Table 3.1.

To reduce current ripple, a symmetric switching pattern is applied. Figure 3.5 illustrates the algorithm flowchart of the proposed modulation strategy for SCSC and ABC. The fourth block in Fig. 3.5 is designed to avoid the narrow pulse problem when $d_r$ and $d_d$ are sufficient large or sufficient small. A simple way is to limit $d_r$ and $d_d$ in the interval $[\varepsilon, 1 - \varepsilon]$, and $\varepsilon$ is a small positive constant specified by designers. $P_i$ ($i \in \{1, 2, ..., 6\}$) is the control signal for the switch $S_i$ ($i \in \{1, 2, ..., 6\}$). And $P_i = 1$ means that the $S_i$ is turned on and $P_i = 0$ indicates that $S_i$ is turned off. For safe current commutation, overlap times are inserted. To further reduce the DC current ripple, the switching sequences of ABC are different under charging and discharging modes.

## 3.3  Selection of the Buffer Capacitor in ABC

Assume that the converter operates at unity input power factor. According to previous analysis, in steady state variables $d_d$ and $d_r$ are expressed as

$$d_r = I \cos(\omega t)/i_{dc} \tag{3.23}$$

$$d_d = u_{dc} \cos(2\omega t)/u_d \tag{3.24}$$

Combining (3.6) with (3.24), the following inequality is obtained.

$$\left| u_{dc} \cos(2\omega t) \Big/ \sqrt{\bar{u}_d^2 + \frac{VI \sin(2\omega t + \varphi)}{2\omega C_d}} \right| \leq 1 \tag{3.25}$$

By neglecting the power losses, $VI$ can be replaced by $2P_o$. According to (3.7) and (3.25), the constraints of $\bar{u}_d$ are expressed as follows:

$$\begin{cases} -u_{dc}^2 \left[ \sin(2\omega t) + \frac{1}{2\omega C_d} \frac{P_o}{u_{dc}^2} \right]^2 + \left( \frac{P_o}{2\omega C_d u_{dc}} \right)^2 + u_{dc}^2 \leq \bar{u}_d^2 \\ \bar{u}_d^2 \geq \frac{P_o}{\omega C_d} \end{cases} \tag{3.26}$$

**Table 3.1**  Duty ratio of each switch

| $u_g$ | $d_1$ | $d_2$ | $d_3$ | $d_4$ | $d_d$ | $d_5$ | $d_6$ |
|-------|-------|-------|-------|-------|-------|-------|-------|
| $u_g > 0$ | 1 | $1 - d_r$ | 0 | $d_r$ | $d_d > 0$ | $1 - d_d$ | 0 |
| $u_g < 0$ | 0 | $-d_r$ | 1 | $1 + d_r$ | $d_d < 0$ | $-d_d$ | 1 |

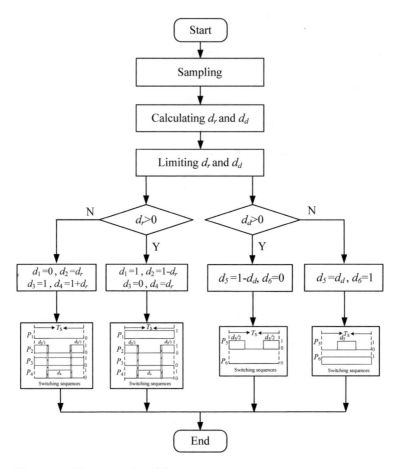

**Fig. 3.5** Flowchart of the proposed modulation strategy

Finally, the average voltage $\bar{u}_d$ should meet

$$\bar{u}_d \geq \begin{cases} \sqrt{\dfrac{P_o}{\omega C_d}} & \dfrac{1}{2\omega C_d} \geq \dfrac{u_{dc}^2}{P_o} \\ \sqrt{u_{dc}^2 + \dfrac{P_o^2}{4\omega^2 C_d^2 u_{dc}^2}} & \dfrac{1}{2\omega C_d} < \dfrac{u_{dc}^2}{P_o} \end{cases} \tag{3.27}$$

Obviously, increasing the capacitance of $C_d$ can decrease the average votlage and then the maximum capacitor voltage. Figure 3.6 shows variation of the maximum capacitor voltage $u_{dmax}$ as a function of the capacitor with the parameters listed in Table 3.2. The selection of the buffer capacitor is a tradeoff between $u_{dmax}$ and the cost of the capacitor. In this study, a capacitor, whose detected capacitance value is 91.8 $\mu$F, is employed. With

**Fig. 3.6** Maximum capacitor voltage versus the capacitance capacity

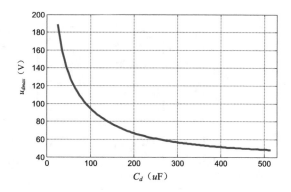

**Table 3.2** Main parameters

| Parameters | Variables | Value |
|---|---|---|
| Amplitude of input phase voltage | $V$ | 92 V |
| Grid angular frequency | $\omega$ | 314 rad/s |
| Input filter inductor | $L_f$ | 0.6 mH |
| Input filter capacitor | $C_f$ | 20 μF |
| DC inductor | $L_{dc}$ | 3 mH |
| DC current reference | $i_{dc}^*$ | 4 A |
| DC side | R load/battery | 8.7 Ω/36 V |
| Switching frequency | $f_s$ | 20 kHz |

proper margin $\bar{u}_d$ is selected to be 80 V, and then the maximum capacitor voltage is 106.4 V according to (3.6).

## 3.4  Simulation and Experimental Results

### 3.4.1  Simulation Results

The presented topology is verified in Matlab/simulink environment. The used parameters are listed in Table 3.2. The simulation results are shown in Fig. 3.7. At the beginning, the decoupling circuit is not activated and the voltage of the decoupling capacitor is zero. Due to a small capacitance of the DC inductor, the DC current $i_{dc}$ is almost rectified sine shape. After the decoupling circuit is activated, the voltage of the decoupling capacitor tracks its reference quickly, and the DC current is a constant with a small fluctuation. As can be observed, the input current is always sinusoidal, and keeps in phase with the input voltage roughly.

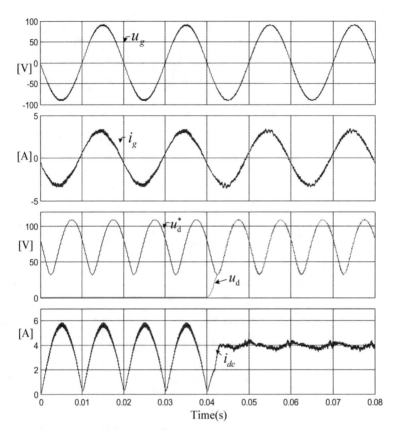

**Fig. 3.7** Simulation waveforms

The power losses are evaluated by the circuit simulator Piece-wise Linear Electrical Circuit Simulation (PLECS). The efficiency curves are illustrated in Fig. 3.8. $\eta$ is the overall conversion efficiency and $\eta_d$ is the efficiency of the added decoupling circuit. Power losses caused by the ABC are around one third of the whole. By power analyzer, the measured efficiency is 84%, which is slightly smaller than the simulation results as the losses caused by the passive components are taken into consideration in experiments.

### 3.4.2  Experimental Results

A prototype for the proposed converter is built in lab for experimental verification. The IGBTs used in the main circuit are 1MBH60D-100, and the control of the converter was realized by a combination of the digital signal processor (DSP) TMS320F28335 and the field programmable gate array (FPGA) EP2C8T144C8N. The voltage of the battery load

**Fig. 3.8** Efficiency curves of the proposed single-phase converter

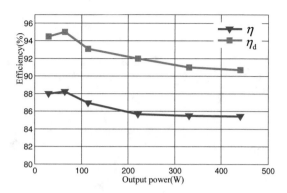

is 36 V, which is composed by three series battery blocks at rated value 12 V/20 AH. To verify the performance of this topology and its related algorithm, two experiments are conducted.

In the first experiment, the load is a resistor of 8.7 Ω. With the proposed decoupling circuit being disabled, the converter works as a conventional SCSC. As illustrated in Fig. 3.9a, the DC current contains a large ripple at twice the frequency of the grid voltage. Once the decoupling circuit is activated, the current ripple is reduced greatly and the DC current is approximately a constant, which are in accord with the simulation results. In the identical situations, if a passive filter is used, it requires a large inductor of 110.8 mH to reach such a low current ripple level. It also can be seen that the input current is sinusoidal and the PFC is 0.97. Harmonic spectrum of the input current is illustrated in Fig. 3.9b. Moreover, Fig. 3.10 shows the THDs of the input current with different switching frequencies. Obviously, increasing switching frequency can improve the input current quality, but it leads to increasing the power losses as well. Under the tradeoff between them, 20 kHz is used in the experiment.

Figure 3.11 shows the spectral analysis of the DC bus current under both conditions mentioned above, where the magnitude is multiplied by ten. Due to effectiveness of the proposed decoupling circuit, the 2rd order component in the DC bus current is reduced to be 12.01% of that in the conventional SCSC. And other low-frequency harmonic components are also much smaller compared with those in the conventional SCSC.

To show the dynamic response of the proposed topology, the experiments with step references were conducted. As shown in Fig. 3.12a, when the DC current reference increases from 2.5 to 4 A abruptly, the DC bus current tracks its reference immediately, the resulting voltage across the buffer capacitor become larger. Figure 3.12b demonstrates the test results of stepping down the DC current reference. In both cases, there is no obvious distortion in the grid current.

Battery loads are widely applied in the electric vehicle, UPS, and so on. Thus the second experiment is conducted to verify the effectiveness of the proposed converter under the battery load condition.

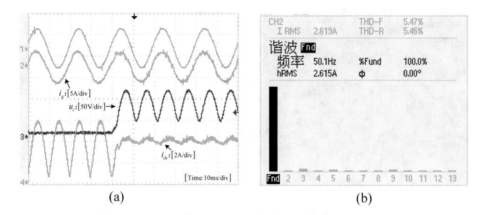

(a)                                          (b)

**Fig. 3.9** Experimental results with resistance load. **a** Experimental waveforms. **b** Spectral analysis
of the steady-state input current with decoupling circuit being activated

**Fig. 3.10**  THDs of the input
current versus the switching
frequency

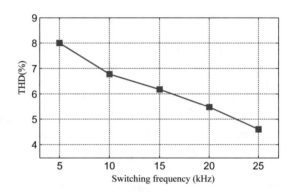

**Fig. 3.11**  Spectral analysis for
the DC bus current

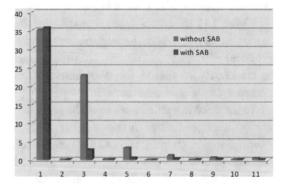

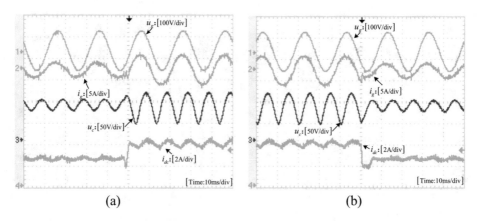

(a)                                                              (b)

**Fig. 3.12** Experimental waveforms with DC current reference changing abruptly **a** from 2.5 to 4 A, **b** from 4 to 2.5 A

Figure 3.13a illustrates the experimental results when charging the battery. As can be seen, $i_{dc}$ is approximately a constant and the input current is sinusoidal. And Fig. 3.13b shows the results of the proposed converter under inversion state. It is clear that the grid current and the grid voltage are phase reversal. Due to complex battery characteristics, the input current with the battery load is worse than that with the resistance load. The THD is 7.2%/7.7% and the PFC is 0.96/0.95 under charging/discharging.

Sometimes converters are required to provide ancillary services such as reactive power and voltage support. Figure 3.14 shows the waveforms when the converter has a power factor of 0.866, i.e. $\varphi = \pm 30°$.

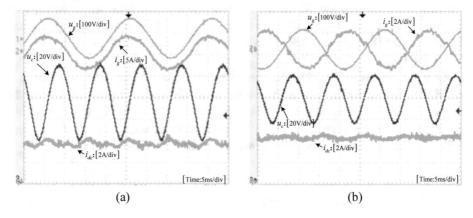

(a)                                                              (b)

**Fig. 3.13** Experimental waveforms with a battery load. **a** Rectification state. **b** Inversion state

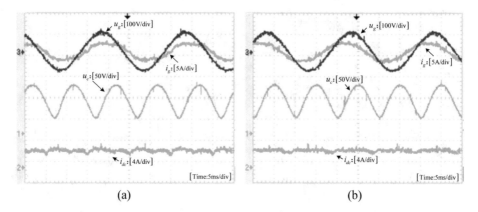

(a)                                              (b)

**Fig. 3.14** Experimental waveforms with $\varphi = \pm 30°$. **a** $\varphi = -30°$, **b** $\varphi = 30°$

## 3.5    Other Independent Decoupling Topologies

Figure 3.15a shows another independent decoupling topology with reducing the usage of the semiconductor devices. It is obtained by replacing the diodes $D_5$ and $D_6$ in Fig. 3.1 with two identical capacitors $C_a$ and $C_b$ ($C_a = C_b$), the buffer capacitor $C_d$ with an inductor $L_1$, and the switch $S_6$ with a diode $D_b$. Consequently, the number of extra switches has been reduced to two, which reduces both the cost and the possibility of switch failures. Moreover, its control is simpler since it only has one active switch. In particular, it employs two buffer capacitors to mitigate the ripple power. Each capacitor only needs to absorb half the ripple power. Thus, its voltage stress is significantly reduced, which avoids the use of high voltage-rated film capacitors. Meanwhile, the low buffer capacitor voltage is beneficial for reducing the DC inductor current ripple. Figure 3.15b shows steady-state experimental waveforms under unity power factor. $v_{com}$ is the sum of the voltages $v_a$ and $v_b$. As seen, the DC-link current is smooth with using a small inductor (5 mH) and good decoupling performance is achieved.

Figure 3.16a shows an AC-DC-AC current source converter with the ABC inserted in the DC bus. It includes three switching bridge arms. The bridge arms A and B form the rectifier, and the bridge arms B and C form the inverter. The main challenge is that the low frequency ripple powers exist both at the source and load sides. Suppose all the ripple powers are buffered by the ABC, the decoupling capacitor voltage is expressed as

**Fig. 3.15** Independent decoupling topology with adding only one active switch and one diode (**a**) and its steady-state experimental waveforms under unity power factor (**b**)

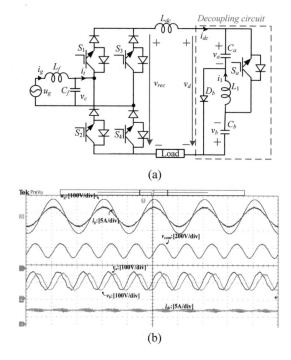

(a)

(b)

$$\begin{cases} u_d = \sqrt{\bar{u}_0^2 + \sqrt{A^2 + B^2 - 2AB\sin(\varphi_i)}\sin(2\omega_i t + \gamma_1) - \sqrt{C^2 + D^2 + 2CD\sin(\varphi_o)}\sin(2\omega_o t + 2\varphi_{io} + \gamma_2)} \\[2mm] A = \frac{V_i I_i}{2\omega_i C_d}, B = \frac{C_g V_i^2}{2C_d}, C = \frac{V_o I_o}{2\omega_o C_d}, D = \frac{C_o V_o^2}{2C_d} \\[2mm] \gamma_1 = \arctan \frac{V_i I_i \sin(\varphi_i) - \omega_i C_g V_i^2}{V_i I_i \cos(\varphi_i)}, \gamma_2 = \arctan \frac{V_o I_o \sin(\varphi_o) + \omega_o C_o V_o^2}{V_o I_o \cos(\varphi_o)} \end{cases}$$

(3.28)

This result is obtained by assuming that

$$u_g = V_i \cos(\omega_i t), i_g = I_i \cos(\omega_i t + \varphi_i)$$
$$u_o = V_o \cos(\omega_o t + \varphi_{io}), i_o = I_o \cos(\omega_o t + \varphi_{io} + \varphi_o)$$

(3.29)

It can be found that the decoupling capacitor voltage highly dependents on the parameters. Therefore, the decoupling performance may be poor if the decoupling control (designed similar to that in Fig. 3.4) is achieved by tracking the given reference $u_d^*$. To achieve good decoupling performance, the control idea that the DC-link current is regulated by the decoupling circuit and the averaged decoupling capacitor voltage is maintained by the rectifier is adopted. The ripple power buffer is automatically achieved. Figure 3.16b, c show the experimental waveforms when the grid frequency is 50 Hz and

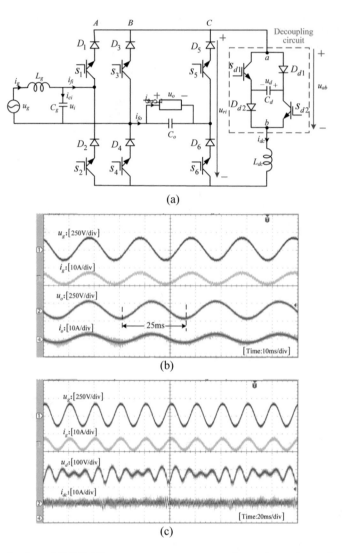

**Fig. 3.16** Single-phase AC-DC-AC current source converter and experimental waveforms when the grid frequency is 50 Hz and the frequency of the out voltage is 40 Hz. **a** Main circuit. **b** Grid voltage/current and output voltage/current. **c** Grid voltage/current, decoupling capacitor voltage, and DC-link current

the output voltage frequency is 40 Hz. It can be found that the with the adopted control, the DC-link current is smooth and good decoupling performance is obtained.

## 3.6 Conclusion

This chapter presents a SCSC with power decoupling capability using an ABC. It works under rectification state as well as inversion state. With the proposed converter, the sinusoidal grid current and low ripple DC current are achieved. The capacitor used in the ABC can be a film capacitor with low rated voltage, which extends life and reduces the size and weight. The proposed decoupling technology has reduced the presence of second harmonic current ripple by 87.99% with the proposed control method where the control reference is the buffer capacitor voltage. The proposed converter is suitable for single-phase rectifiers, UPS, V2G and the PV generation system. The validity of the proposed converter and control strategy was confirmed experimentally.

## Bibliography

1. Bush, C. R., & Wang, B. (2009). A single-phase current source solar inverter with reduced-size DC-link. In *Proceedings of the IEEE Energy Conversion Congress and Exposition (ECCE)*, San Jose, CA (pp. 54–59).
2. Chaudhary, P., & Sensarma, P. (2014). Front-end buck rectifier with reduced filter size and single-loop control. *IEEE Transactions on Industrial Electronics, 60*(10), 4359–4368.
3. Yuan, H., Li, S., Tan, S., & Hui, S. Y. R. (2020). Internal dynamics stabilization of single-phase power converters with Lyapunov-based automatic-power-decoupling control. *IEEE Transactions on Power Electronics, 35*(2), 2160–2169.
4. Hashimoto, T., & Sone, S. (1992). Single-phase PWM converter using balanced two-phase rectification. *Electrical Engineering in Japan, 112*(3), 215–220.
5. Han, H., Liu, Y., Sun, Y., Su, M., & Xiong, W. (2015). Single-phase current source converter with power decoupling capability using a series-connected active buffer. *IET Power Electronics, 8*(5), 700–707.
6. Sanders, J. A., Verhulst, F., & Murdock, J. (2007). *Averaging methods in nonlinear dynamical systems*. Springer.
7. Vitorino, M. A., Hartmann, L. V., Fernandes, D. A., Silva, E. L., & Corrêa, M. B. R. (2014). Single-phase current source converter with new modulation approach and power decoupling. In *Proceedings of the IEEE APEC*, Fort Worth, TX (pp. 2200–2207).
8. Nonaka, S., & Neba, Y. (1993). Single-phase PWM current source converter with double-frequency parallel resonance circuit for DC smoothing. *IEEE IAS Annual Meeting Record* (pp. 1144–1151).
9. Ohnuma, Y., & Itoh, J. I. (2014). A novel single-phase buck PFC AC-DC converter with power decoupling capability using an active buffer. *IEEE Transactions on Industry Applications, 50*(3), 1905–1914.
10. Oruganti, R., & Palaniapan, M. (2000). Inductor voltage control of buck-type single-phase AC–DC converter. *IEEE Transactions on Power Electronics, 15*(2), 411–416.
11. Su, M., Long, X., Sun, Y., et al. (2014). An active power decoupling method for single-phase AC-DC converters. *IEEE Transactions on Industrial Informatics, 10*(1), 461–468.
12. Tang, Y., Zhu, D., Jin, C., et al. (2014). A three-level quasi-two-stage single-phase PFC converter with flexible output voltage and improved conversion efficiency. *IEEE Transactions on Industrial Electronics*

13. Wang, H., Chung, H. H., & Liu, W. (2014). Use of a series voltage compensator for reduction of the DC-link capacitance in a capacitor supported system. *IEEE Transactions on Power Electronics, 29*(3), 1163–1175.
14. Liu, Y., Tang, S., Wang, H., Ning, G., & Xiong, W. (2021). Independent power decoupling method using minimum switch devices for single-phase current source converters. *Journal of Power Electronics, 21*(9), 1383–1394.
15. Liu, Y., Sun, Y., Su, M., Li, X., & Ning, S. (2018). A single-phase AC-DC-AC converter with unified ripple power decoupling. *IEEE Transactions on Power Electronics, 33*(4), 3204–3217.

# Dependent Decoupling Topologies for Current-Source Rectifier

4

In last chapter the introduced independent decoupling topology involves a lot of additional semiconductor devices (four semiconductor devices), which is adverse to cost and efficiency. That is also a major shortcoming of the dependent decoupling topology. In this chapter, switching multiplexing decoupling circuits are studied, including the double capacitors decoupling circuit and the single capacitor decoupling circuit. Both of them require no additional active switching devices. In this chapter, the topological structure, working principle, switching state, modulation strategy, and control method of the two proposed decoupling circuits are explained in Sects. 4.1 and 4.2. The decoupling capacitor voltage is analyzed and compared in Sect. 4.3. Section 4.4 demonstrates the simulation and experimental study. Some other dependent decoupling topologies are introduced and discussed in Sect. 4.5.

## 4.1    Switch-Multiplexing Circuit Topology with Double Decoupling Capacitors

### 4.1.1    Circuit Configuration

The proposed switch-multiplexing circuit topology with double decoupling capacitors and its derivation process are shown in Fig. 4.1a and b, respectively. Compared with the conventional SCSC, the proposal requires two additional identical capacitors to buffer the ripple power ($C_1$ and $C_2$) and two more diodes ($D_3$ and $D_4$) to guarantee the safety operation of the converter during start-up and stop processes.

© The Author(s), under exclusive license to Springer Nature Switzerland AG 2023
Y. Liu, *Active Power Decoupling Technology in Single-Phase Current-Source Converters*,
Synthesis Lectures on Power Electronics, https://doi.org/10.1007/978-3-031-21270-3_4

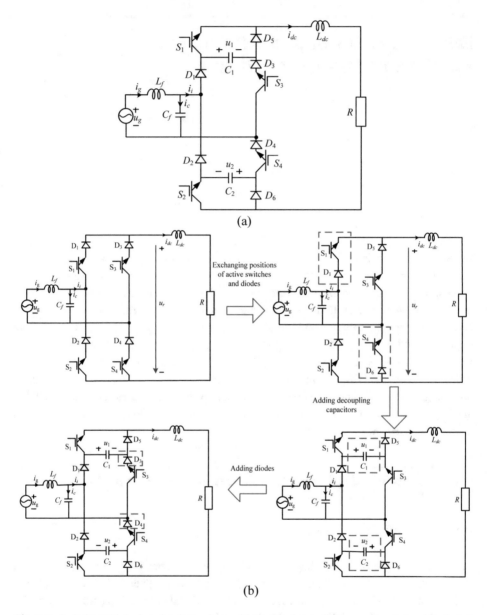

**Fig. 4.1** Switch-multiplexing circuit topology with double decoupling capacitors (**a**) and its derivation process (**b**)

## 4.1.2  Switching States

In conventional single-phase SCSC, each switching state involves two active switches and only four switching states are available. But in the proposed SCSC the number of the conduction switches in each switching state can be one, two, or three, which enriches the available switching states greatly. To avoid misgating-on of diodes $D_1$ and $D_2$, the decoupling capacitor voltages $u_1$ and $u_2$ are always controlled to be higher than $|u_g|$. Then, there are eleven available switching states, which are shown in Fig. 4.2. In the figures, $T = (S_1 \ S_2 \ S_3 \ S_4)$ denotes the states of the four active switches, where $S_i = $ '1' $(i = 1, 2, 3, 4)$ indicates the corresponding switch $S_i$ is turned on and '0' turned off.

The switching states are further divided into four groups in terms of functions. The first group includes switching states 1 and 2, in which the ac side is connected into the dc-loop and the decoupling capacitors $C_1$ and $C_2$ are both bypassed. This group is exclusively used for synthesizing the input current. The second group consists of switching states 3, 4, 5, and 6, in which the ac side is disconnected from the dc-loop on the right side. This group is exclusively used for decoupling the ripple power. The third group is composed of switching states 7, 8, 9, and 10, in which the ac side capacitor $C_f$ and one of the decoupling capacitor are connected in series to supply the DC loads. This group is carried out to achieve the goal of input current synthesis and ripple power buffering simultaneously. The forth group is composed of the switching state 11. This group provides the freewheeling path for the DC-link current $i_{dc}$.

According to Fig. 4.2, the equivalent circuit model of the proposed converter is illustrated in Fig. 4.3. And $u_r$ is the average value of the rectified output voltage over a switching period in the conventional single-phase SCSC. The equivalent series voltage $u_s$ is provided by controlling the decoupling part. If the ripple power is completely buffered by the decoupling capacitors $C_1$ and $C_2$, $u_s$ should be

$$u_s = \frac{P_r}{i_{dc}} = \frac{VI \cos(2\omega t + \varphi)}{2i_{dc}} \tag{4.1}$$

When $u_s$ is positive/negative, the decoupling capacitors absorb/release ripple energy. From the view of voltage-second balance, $u_s$ is used to ensure that the net voltage-second of the DC-link inductor $L_{dc}$ is zero over each switching period $T_s$. Then only a small DC-link inductor is required to filter high frequency harmonics caused by switches.

## 4.1.3  Modulation Scheme

To accomplish the dual purposes of input current synthesis and ripple power mitigation, a hybrid modulation method is developed. During each line frequency period ac current reference $i_{i\_ref}$, which is the average value over a switching period $T_s$ and synthesized by the DC-link current $i_{dc}$, is classified as two main sectors: $i_{i\_ref} > 0$ and $i_{i\_ref} \leq 0$. And

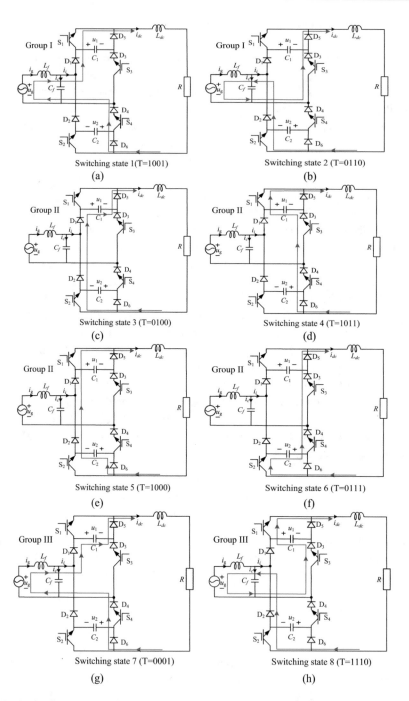

**Fig. 4.2** Switching states

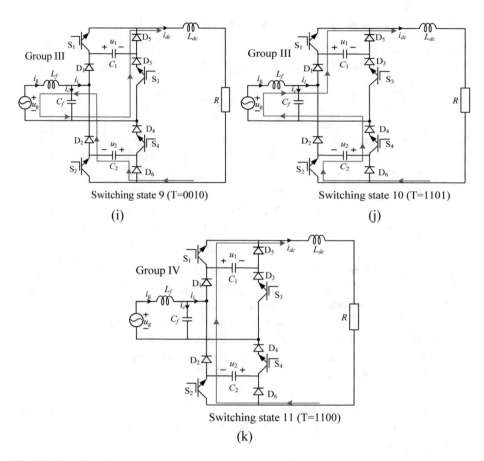

Switching state 9 (T=0010)

(i)

Switching state 10 (T=1101)

(j)

Switching state 11 (T=1100)

(k)

**Fig. 4.2** (continued)

**Fig. 4.3** Equivalent circuit model of the double capacitors decoupling circuit

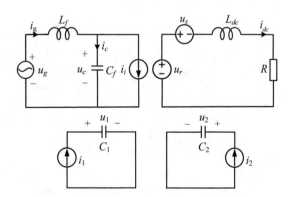

**Table 4.1** Operation modes and corresponding switching states

| Operation modes | Mode I $C_1$ charging | Mode II $C_2$ discharging | | Mode III $C_2$ charging | Mode IV $C_1$ discharging | | Duty ratios |
|---|---|---|---|---|---|---|---|
| | $i_{i\_ref} > 0$ | | | $i_{i\_ref} \leq 0$ | | | |
| | $u_{s\_ref} > 0$ | $u_{s\_ref} \leq 0$ | | $u_{s\_ref} > 0$ | $u_{s\_ref} \leq 0$ | | |
| | | $T_d = 0$ | $T_d = 1$ | | $T_d = 0$ | $T_d = 1$ | |
| Switching sates (T) | 7 (T = 0001) | 10 (T = 1101) | 1 (T = 1001) | 9 (T = 0010) | 8 (T = 1110) | 2 (T = 0110) | $d_1$ |
| | 1 (T = 1001) | 6 (T = 0111) | 6 (T = 0111) | 2 (T = 0110) | 4 (T = 1011) | 4 (T = 1011) | $d_2$ |
| | 11 (T = 1100) | 1 (T = 1001) | 11 (T = 1100) | 11 (T = 1100) | 2 (T = 0110) | 11 (T = 1100) | $d_3$ |

each main sector is further divided into two modes according to the operations of decoupling capacitors, i.e., charging operation ($u_{s\_ref} > 0$) and discharging operation ($u_{s\_ref} \leq 0$). Therefore, there are four basic operational modes and three possible switching states exist in each mode, which are summarized in Table 4.1. In discharging modes, if the sum of the durations of synthesizing input current and decoupling is more than unity, $T_d = 0$; or $T_d = 1$. $d_1$, $d_2$, and $d_3$ are duty ratios of the corresponding operational state and $d_1 + d_2 + d_3 = 1$. To reduce the DC-link current ripple, the distributions of $d_1$, $d_2$, and $d_3$ in each switching period $T_s$ are shown in Fig. 4.4.

According to the developed hybrid modulation method, steady-state operating waveforms for the proposed single-phase SCSC are shown in Fig. 4.5. Four operational modes are briefly introduced as follows.

(1) *Energy Absorbed Modes*

Modes I and III are energy absorbed modes, in which $u_{s\_ref}$ is always positive. In Mode I, ac current reference $i_{i\_ref}$ is positive and the capacitor $C_1$ is charged; whereas in Mode III ac current reference $i_{i\_ref}$ is negative and the capacitor $C_2$ is charged. Here Mode I is taken for an example to analyze the operational case. According to Fig. 4.4a, in each switching period $T_s$ there are three switching states:

Interval 0 ($0 \leq t < d_1 T_s$): Switching state 7 in Group III is carried out. Then switch $S_4$ is turned on and switches $S_1$, $S_2$, and $S_3$ are turned off. The power in ac grid is transferred to the capacitor $C_1$ and the load.

Interval 1 [$d_1 T_s \leq t < (d_1 + d_2)T_s$]: Switching state 1 in Group II is carried out. In this state switches $S_1$ and $S_4$ are turned on and $S_2$ and $S_3$ are turned off. In this state the decoupling capacitors are bypassed and the power in the ac side is transferred to the load, which is identical to that in the conventional SCSC.

**Fig. 4.4**
Distributions of $d_1$,
$d_2$, and $d_3$ in one switching
period $T_s$. **a** Charging.
**b** Discharging when $T_d = 0$.
**c** Discharging when $T_d = 1$

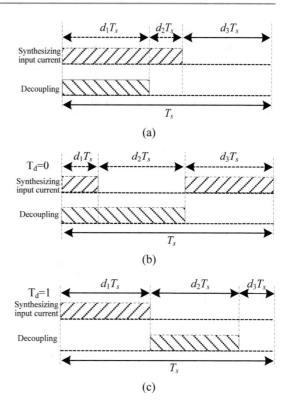

(a)

(b)

(c)

Interval 2 [$(d_1 + d_2)T_s \leq t \leq T_s$]: Switching state 11 in Group IV is carried out. Then switches $S_1$ and $S_2$ are turned on and $S_3$ and $S_4$ are turned off. Both of the capacitors $C_1$ and $C_2$ and ac grid are bypassed. Here, switching state 11 ($T = 1100$) is adopted rather than $T = 0011$ to reduce the conduction power losses. In other modes the freewheeling is also accomplished by employing switching state 11 for the same purpose.

*(2) Energy Released Modes*

Modes II and IV are energy released modes, in which $u_{s\_ref}$ is always negative. In Mode II ac current reference $i_{i\_ref}$ is positive, and the capacitor $C_2$ is discharged; whereas in Mode IV ac current reference $i_{i\_ref}$ is negative and the capacitor $C_1$ is discharged. In each discharging mode there are two subintervals, i.e., Sub-I and Sub-II. That is, the absorbed energy will be released by twice. In the following, Mode IV is taken for an example to analyze the operational case. According to Fig. 4.4b, c, in each switching period $T_s$ there are three switching states:

Interval 0 ($0 \leq t < d_1 T_s$): If $T_d = 0$, switching state 8 in Group III is carried out. Then switches $S_1$, $S_2$, and $S_3$ are turned on and $S_4$ are turned off. The power from ac grid and the capacitor $C_1$ is transferred to the load. If $T_d = 1$, switching state 2 in Group I is carried out. In this state switches $S_2$ and $S_3$ are turned on and $S_1$ and $S_4$ are turned off. This state is identical to that in the conventional SCSC.

Interval 1 $[d_1 T_s \le t < (d_1 + d_2) T_s]$: Switching state 4 in Group II is carried out in despite of the value of $T_d$. In this state switches $S_1$, $S_3$, and $S_4$ are turned on and $S_2$ is turned off. This state contributes to discharging capacitor $C_1$.

Interval 2 $[(d_1 + d_2) T_s \le t \le T_s]$: If $T_d = 0$, switching state 2 is carried out again, i.e., the duration of synthesizing input current is split into two pieces in a switching period. If $T_d = 1$, switching state 11 is carried out for freewheeling.

(3) *Expressions of Decoupling Capacitor Voltages*

As for the capacitor $C_1$, as can be seen from Fig. 4.5, in Mode I it is charged to absorb the ripple power. Therefore, the following differential equation is satisfied,

$$C\frac{du_1}{dt}u_1 = P_r \tag{4.2}$$

where $u_1$ is the voltage of capacitors $C_1$. By integrating both sides of equation (4.2) with respect to time, $u_1$ can be expressed as

$$u_1 = \sqrt{\bar{u}_0^2 + \frac{VI\sin(2\omega t + \varphi)}{2\omega C}} \quad \text{Mode I} \tag{4.3}$$

where $\bar{u}_0$ is the DC component of $u_1$.

In Modes II and III, capacitor $C_1$ is bypassed. Then $u_1$ is kept constant and expressed as

$$u_1 = \begin{cases} \sqrt{\bar{u}_0^2 - \dfrac{VI}{2\omega C}} & \text{Mode II-Sub-I} \\[3mm] \sqrt{\bar{u}_0^2 + \dfrac{VI}{2\omega C}} & \text{Mode II-Sub-II} \\[3mm] \sqrt{\bar{u}_0^2 - \dfrac{VI\sin(\varphi)}{2\omega C}} & \text{Mode III} \end{cases} \tag{4.4}$$

In Mode IV, the capacitor $C_1$ is discharged to release the ripple power. As the ripple energy is released by twice, $u_1$ is expressed as

$$u_1 = \begin{cases} \sqrt{\bar{u}_0^2 - \dfrac{VI\cos(2\omega t + 2\varphi)}{2\omega C}} & \text{Mode IV-Sub-I} \\[3mm] \sqrt{\bar{u}_0^2 + \dfrac{VI\cos(2\omega t + 2\varphi)}{2\omega C}} & \text{Mode IV-Sub-II} \end{cases} \tag{4.5}$$

Similar analysis can be done on $C_2$. Then $u_2$ is expressed as

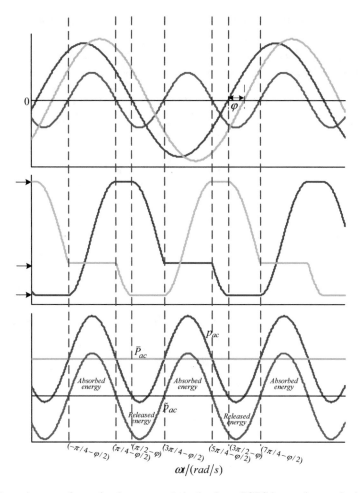

**Fig. 4.5** Operating waveforms for the proposed single-phase SCSC in steady state by ignoring the effect of input filters

$$
u_2 = \begin{cases}
\sqrt{\bar{u}_0^2 - \dfrac{V I \sin(\varphi)}{2\omega C}} & \text{Mode I} \\[2ex]
\sqrt{\bar{u}_0^2 - \dfrac{V I \cos(2\omega t + 2\varphi)}{2\omega C}} & \text{Mode II-Sub-I} \\[2ex]
\sqrt{\bar{u}_0^2 + \dfrac{V I \cos(2\omega t + 2\varphi)}{2\omega C}} & \text{Mode II-Sub-II} \\[2ex]
\sqrt{\bar{u}_0^2 + \dfrac{V I \sin(2\omega t + \varphi)}{2\omega C}} & \text{Mode III} \\[2ex]
\sqrt{\bar{u}_0^2 - \dfrac{V I}{2\omega C}} & \text{Mode IV-Sub-I} \\[2ex]
\sqrt{\bar{u}_0^2 + \dfrac{V I}{2\omega C}} & \text{Mode IV-Sub-II}
\end{cases}
\tag{4.6}
$$

Apparently, $u_1$ and $u_2$ are both piecewise functions. Moreover, $u_1$ can be obtained by shifting $u_2$ to the right with $\pi$ rad/s. It is also founded that for each decoupling capacitor the discharging durations of Sub-I and Sub-II are respectively $(\pi/4 - \varphi/2)/\omega$ and $(\pi/4 + \varphi/2)/\omega$ during each line frequency period.

According to the above-mentioned analysis and Fig. 4.4, the synthesized current $i_i$ and the equivalent voltage $u_s$ provided by $C_1$ or $C_2$ are obtained as follows:

$$
i_i = \begin{cases} \mathrm{sgn}(i_{i\_ref})(d_1 + d_2)i_{dc} & \text{Mode I or III} \\ \mathrm{sgn}(i_{i\_ref})[d_1 + (1 - T_d)d_3]i_{dc} & \text{Mode II or IV} \end{cases} \tag{4.7}
$$

$$
u_s = \begin{cases} d_1 u_1 & \text{Mode I} \\ [d_2 + (1 - T_d)d_1]u_2 & \text{Mode II} \\ d_1 u_2 & \text{Mode III} \\ [d_2 + (1 - T_d)d_1]u_1 & \text{Mode IV} \end{cases} \tag{4.8}
$$

where sgn () is the sign function.

### 4.1.4 Modeling and Control

According to Fig. 4.3, the average model of the proposed converter is formulated as follows:

$$
L_f \frac{di_g}{dt} = u_g - u_c \tag{4.9}
$$

$$
C_f \frac{du_c}{dt} = i_g - i_i \tag{4.10}
$$

$$
L_{dc} \frac{di_{dc}}{dt} = \frac{i_i}{i_{dc}} u_c - u_s - R i_{dc} \tag{4.11}
$$

$$
C_1 \frac{du_1}{dt} = \begin{cases} \frac{u_s}{u_1} i_{dc}, & i_i \cdot u_s \geq 0 \\ 0, & i_i \cdot u_s < 0 \end{cases} \tag{4.12}
$$

$$
C_2 \frac{du_2}{dt} = \begin{cases} \frac{u_s}{u_2} i_{dc}, & i_i \cdot u_s < 0 \\ 0, & i_i \cdot u_s \geq 0 \end{cases} \tag{4.13}
$$

where $i_g$, $i_{dc}$, $u_1$, and $u_2$ are controlled variables; $i_i$ and $u_s$ are control input variables.

There are three control targets: (1) to keep DC-link current constant (i.e., buffering the ripple power), (2) to obtain sine input current, (3) to maintain DC component of $u_1$ and $u_2$ (i.e., $\bar{u}_0$). According to (4.11), both $i_i$ and $u_s$ could be used for regulating the DC-link current $i_{dc}$ in theory. To improve the overall control performance, $u_s$ is selected and $i_i$ is in charge of implementing PFC as well as maintaining the DC component of the $u_1(u_2)$ at a given level.

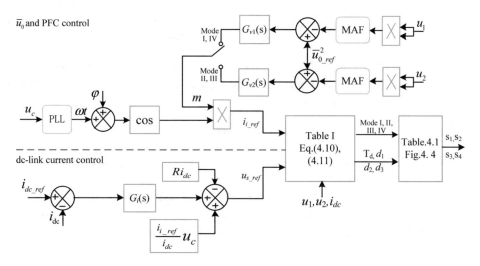

**Fig. 4.6**  Block diagram of the overall control scheme

The overall control block diagram is shown in Fig. 4.6. Regarding to the DC-link current control, a proportional-integral (PI) controller is used. The transfer function of the current controller is expressed as follows:

$$G_i(s) = k_p + \frac{k_i}{s} \tag{4.14}$$

where $k_p$ is the proportional term and $k_i$ is the integral gain. Substitute $u_{s\_ref}$ to (4.11), the closed-loop error dynamic equation for the DC-link current is expressed as

$$L_{dc}\frac{de}{dt} = k_p e + k_i \int e(\tau)d\tau \tag{4.15}$$

where $e = i_{dc\_ref} - i_{dc}$. Clearly, to guarantee the stability, the parameters of $G_i(s)$ should satisfy $k_p < 0$, $k_i < 0$. In addition, the bandwidth of the DC-link current subsystem is determined by $k_i$, and the damping ratio is dependent on $k_p$. In both simulations and experiments, the control bandwidth is designed to be around 200 Hz and the damping ratio is selected to be 0.707 for fast dynamic responses in DC-link current control.

Before introducing the voltage control of $C_1$ and $C_2$, assume the DC-link current subsystem is in steady state, according to (4.11), $u_s$ in steady state can be expressed as

$$u_s = \frac{i_i}{i_{dc\_ref}}u_c - Ri_{dc\_ref} \tag{4.16}$$

Substitute (4.16) to (4.12), the obtained voltage dynamic of $C_1$

$$\frac{C_1}{2}\frac{d\,x}{d\,t} = \begin{cases} u_c i_i - Ri^2_{dc\_ref}, & i_i u_s \geq 0 \\ 0, & i_i u_s < 0 \end{cases}$$ (4.17)

where $x = u_1^2$. Clearly, the voltage across $C_1$ can be controlled by control input $i_i$.

To obtain a sinusoidal grid current, the average of $i_i$ over a switching period should satisfy the following expression:

$$i_i = m \cdot \cos(\theta + \varphi)$$ (4.18)

where $\theta$ is the phase of $u_c$, which is obtained by a digital PLL. $m$ is the control input, which will be designed latter.

If the subsystem (4.9) and (4.10) are stable and the effect of the input filter can be ignored, $u_c$ could be approximated to $V\cos(\theta)$. As (4.17) is a periodic system, the periodic averaging method is used to design the controller. Then the average differential equation of (4.17) is

$$C_1\frac{d\overline{x}}{d\,t} = 0.5mV\cos(\varphi) - Ri^2_{dc\_ref}$$ (4.19)

where $\overline{x} = \frac{1}{T}\int_{t-T}^{t} x(\tau)d\tau$, and $\overline{x}$ is obtained by a moving average filter (MAF) in implementation. Then, $G_{v1}(s)$ is designed as

$$G_{v1}(s) = k_{p1} + \frac{k_{i1}}{s}$$ (4.20)

For stability, the parameters of $G_{v1}(s)$ should satisfy $k_{p1} > 0$, $k_{i1} > 0$. Since $C_1$ and $C_2$ are equivalent in functionalities, $G_{v2}(s)$ is designed to be same with $G_{v1}(s)$. In the decoupling capacitor voltage control loop, to make a trade-off between the input current quality and the decoupling capacitor voltage control, the designed control bandwidth is around 10 Hz and the selected damping ratio is 2.0 in the subsequent simulations and experiments.

After the references of $u_s$ and $i_i$ have been determined, by combining (4.7) and (4.8), the duty ratios for all the switches can be calculated.

## 4.2    Switch-Multiplexing Circuit Topology with Single Decoupling Capacitor

### 4.2.1    Circuit Configuration

Figure 4.7 shows the switch-multiplexing circuit topology with single decoupling capacitor. It is formed by removing a decoupling capacitor and a diode in the upper switching arm from the topology in Fig. 4.1a. Consequently, the added hardware is reduced. In this situation, the second ripple power is buffered only by $C_d$.

**Fig. 4.7** Switch-multiplexing circuit topology with single decoupling capacitor

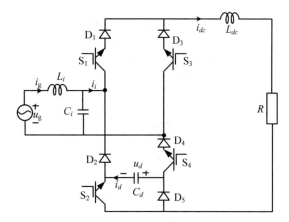

### 4.2.2   Switching States

In Fig. 4.7, $S_1$ and $S_3$ are complementary, whereas $S_2$ and $S_4$ do not need to obey that constraint. Figure 4.8 illustrates six switching states, which are used in this chapter. The switching states are divided into three groups in terms of functions. The first group includes the switching states 1 and 2, in which the ac side is connected to the DC-link current loop and the decoupling capacitor $C_d$ is bypassed. This group contributes to synthesizing the input current (transferring active power). The second group consists of switching states 3 and 4, in which the ac side is by passed and the decoupling capacitor $C_d$ works. This group is exclusively used to synthesize the decoupling capacitor current (buffer the ripple power). The third group is composed of switching states 5 and 6, which provide a freewheeling path for the DC-link current $i_{dc}$. The effects of different switching states on currents $i_i$ and $i_d$ are summarized in Table 4.2. And the expected currents $i_i$ and $i_d$ can be synthesized by using the six switching states.

### 4.2.3   Modulation Scheme

By ignoring the power losses, similar to the (4.3), the decoupling capacitor voltage $u_d$ and current $i_d$ can be expressed as:

$$u_d = \sqrt{\bar{u}_0^2 + \frac{VI\sin(2\omega t + \varphi)}{2\omega C_d}} \tag{4.21}$$

$$i_d = \frac{VI\cos(2\omega t + \varphi)/2}{\sqrt{\bar{u}_0^2 + \frac{VI\sin(2\omega t+\varphi)}{2\omega C_d}}} \tag{4.22}$$

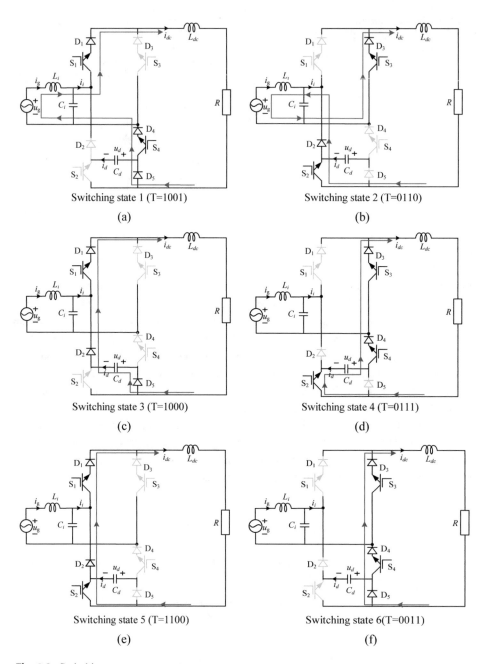

**Fig. 4.8** Switching states

**Table 4.2** Switching states and currents

| Switching states | $i_i$ | $i_d$ |
|---|---|---|
| 1 | $i_{dc}$ | 0 |
| 2 | $-i_{dc}$ | 0 |
| 3 | 0 | $i_{dc}$ |
| 4 | 0 | $-i_{dc}$ |
| 5,6 | 0 | 0 |

Assume that $d_j$ ($j = 1, 2, 3, 4, 5, 6$) is the duty ratio of the switching state $j$, six duty ratios are subject to the following equation:

$$\sum_{j=1}^{6} d_j = 1 \tag{4.23}$$

Then, $i_i$ and $i_d$ can be expressed as:

$$\begin{bmatrix} i_i \\ i_d \end{bmatrix} = \begin{bmatrix} d_1 - d_2 \\ d_3 - d_4 \end{bmatrix} i_{dc} \tag{4.24}$$

Then, $d_j$ can be expressed as:

$$\begin{cases} d_1 = \begin{cases} i_{i\_ref}/i_{dc} & i_{i\_ref} > 0 \\ 0 & i_{i\_ref} \leq 0 \end{cases} \quad d_2 = \begin{cases} 0 & i_{i\_ref} > 0 \\ -i_{i\_ref}/i_{dc} & i_{i\_ref} \leq 0 \end{cases} \\ d_3 = \begin{cases} i_{d\_ref}/i_{dc} & i_{d\_ref} > 0 \\ 0 & i_{d\_ref} \leq 0 \end{cases} \quad d_4 = \begin{cases} 0 & i_{d\_ref} > 0 \\ -i_{d\_ref}/i_{dc} & i_{d\_ref} \leq 0 \end{cases} \\ d_5 = \begin{cases} 1 - \sum_{j=1}^{4} d_j & i_{d\_ref} > 0 \\ 0 & i_{d\_ref} \leq 0 \end{cases} \quad d_6 = \begin{cases} 0 & i_{d\_ref} > 0 \\ 1 - \sum_{j=1}^{4} d_j & i_{d\_ref} \leq 0 \end{cases} \end{cases} \tag{4.25}$$

where $i_{i\_ref}$ and $i_{d\_ref}$ are the references of the grid current and decoupling capacitor current, respectively. To reduce the DC-link current ripple, in each switching period $T_s$, the decoupling operation is firstly carried out when charging $C_d$; while the power transmission is prior when discharging $C_d$. To balance heat dissipation of each bridge, switching state 5/6 is carried out when charging/discharging $C_d$. Figure 4.9 shows the switching patterns. From Eq. (4.25) and Fig. 4.21, it is easy for the modulation to implement in FPGA (CPLD).

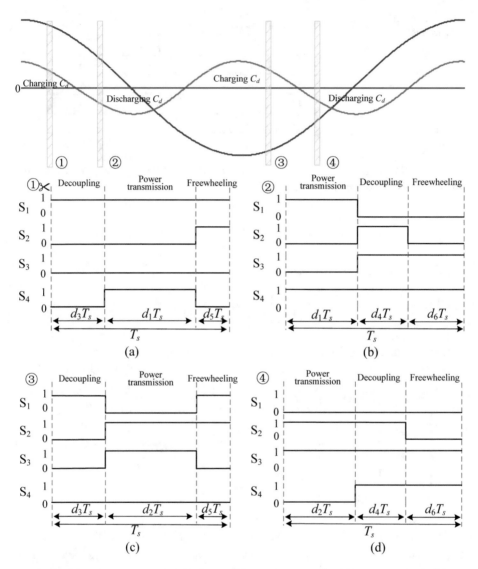

**Fig. 4.9** Switching patterns (① Charging $C_d$ when ac current is positive. ② Discharging $C_d$ when ac current is positive. ③ Charging $C_d$ when ac current is negative. ④ Discharging $C_d$ when ac current is negative)

### 4.2.4 Control Strategy

According to Fig. 4.7, the equivalent averaged circuit model of the proposed converter is illustrated in Fig. 4.10. The averaged voltages $u_r$ and $u_s$ are provided by filter capacitor voltage and decoupling capacitor voltage, respectively. Then the differential equations of the converter are obtained as follows

$$L_f \frac{di_g}{dt} = u_g - u_c \tag{4.26}$$

$$C_f \frac{du_c}{dt} = i_g - i_i \tag{4.27}$$

$$L_{dc} \frac{di_{dc}}{dt} = u_r - u_s - R i_{dc} \tag{4.28}$$

$$C_d \frac{du_d}{dt} = i_d \tag{4.29}$$

$$i_i = (d_1 - d_2) i_{dc} \tag{4.30}$$

$$i_d = (d_3 - d_4) i_{dc} \tag{4.31}$$

$$u_r = \frac{i_i}{i_{dc}} u_c = (d_1 - d_2) u_c \tag{4.32}$$

$$u_s = \frac{i_d}{i_{dc}} u_d = (d_3 - d_4) u_d \tag{4.33}$$

where $i_g$, $i_{dc}$, $u_d$ are output variables; $i_i$ and $i_d$ are control input variables.

The precise open-loop reference (the decoupling capacitor voltage $u_d$ or current $i_d$) for power decoupling is difficult to obtain due to power losses and parameter perturbations. To achieve good decoupling performance a closed-control strategy is adopted. Its basic idea is that $i_d$ is responsive for regulating DC-link current $i_{dc}$ and $i_i$ is in charge of PFC as well as maintaining the dc component of the $u_d$ at a given level.

(1) *Ripple Power Control*

Actually, if the ripple power is not completely absorbed by the decoupling capacitor $C_d$, the residual part will be imposed on the DC-link inductor $L_{dc}$. Then the voltage-second

**Fig. 4.10** Equivalent averaged circuit of the single capacitor decoupling circuit

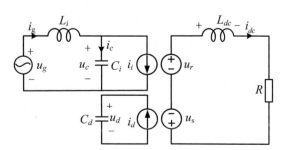

balance will be broken and the DC-link current cannot be kept constant. Therefore, the error between the DC-link current reference and its detected value can be used to reflect the decoupling effect indirectly. As shown in Fig. 4.11, the error is sent to a PI controller. Then the reference $i_{d\_ref}$ can be obtained as follows:

$$i_{d\_ref} = i_d + \frac{i_{dc}}{u_d} G_i(s)(i_{dc\_ref} - i_{dc}) \tag{4.34}$$

where $i_d$ serves as a forward compensation in the control to improve the dynamic response.

### (2) Decoupling Capacitor Voltage Control

The decoupling capacitor voltage fluctuates as a result of buffering the ripple power. Though the required average ripple power through $C_d$ over a line frequency cycle is zero, the power losses due to switches and capacitors are unavoidable. Therefore, the dc component of the decoupling capacitor voltage should be maintained at a predetermined level. Before introducing the voltage control of $C_d$, assume the DC-link current subsystem is in steady state.

According to (4.28) and (4.32), $u_s$ in steady state can be expressed as

$$u_s = \frac{i_i}{i_{dc}} u_c - R i_{dc} \tag{4.35}$$

Substitute (4.32) and (4.35) to (4.29), the voltage dynamic of $C_d$ is expressed as follows,

$$\frac{C_d}{2} \frac{dx}{dt} = u_c i_i - R i_{dc}^2 \tag{4.36}$$

where $x = u_d^2$. Clearly, the voltage across $C_d$ can be controlled by control input $i_i$. By ignoring the effects of the input filters, $i_i \approx i_g$ and $u_c \approx u_g$, then (4.36) can be rewritten

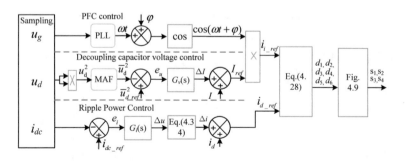

**Fig. 4.11** Block diagram of the control scheme

as

$$\frac{C_d}{2}\frac{dx}{dt} = \frac{VI}{2}[\cos(\varphi) + \cos(2\omega t + \varphi)] - Ri_{dc}^2 \tag{4.37}$$

Since (4.37) is a periodic system, the periodic averaging method is used to design the controller. Then the average differential equation of (4.37) is

$$C_d\frac{d\overline{x}}{dt} = VI\cos(\varphi) - 2Ri_{dc}^2 \tag{4.38}$$

where $\overline{x} = \overline{u}_d^2 = \frac{1}{T}\int_{t-T}^{t} x(\tau)d\tau$, and $\overline{x}$ is obtained by a moving average filter (MAF) in implementation. Equation (4.38) is a simple first-order system, thus $G_v(s)$ is achieved by a PI controller. Then the reference $i_{i\_ref}$ can be written as

$$i_{i\_ref} = I\_ref \cos(\omega t + \varphi) = \left\{I + G_v(s)\left(\overline{u}_{d\_ref}^2 - \overline{u}_d^2\right)\right\}\cos(\omega t + \varphi) \tag{4.39}$$

where $I = 2\,\overline{P}/V$ is the feed-forward term to increase the dynamic response of the current control.

Figure 4.11 shows the overall block diagram of the control scheme. It mainly includes a phase locking loop (PLL), a voltage controller and a current controller. Compared with the control scheme, the complexity does not increase a lot. And a highlight merit of the proposed control is the feed-back regulation of the DC-link current, which enhances the decoupling effects.

## 4.3   Voltage Stress Analysis

As for the switch-multiplexing circuit topology with double decoupling capacitors, according to the analysis in Sect. 4.1, the decoupling capacitor voltage should be less than the device stress $V_M$ and greater than the input voltage. According to the former constraint, we have

$$\sqrt{\overline{u}_0^2 + \frac{VI\sin(2\omega t + \varphi)}{2\omega C_d}} \leq V_M \tag{4.40}$$

Here the value of the $C_1$ and $C_2$ is supposed to be $C_d$. Then, $\overline{u}_0$ should meet

$$\overline{u}_0 \leq \sqrt{V_M^2 - \frac{VI}{2\omega C_d}} \tag{4.41}$$

According to the second constraint, we have

$$\sqrt{\bar{u}_0^2 + \frac{VI \sin(2\omega t + \varphi)}{2\omega C_d}} \geq |V\cos(\omega t)| \tag{4.42}$$

Then, $\bar{u}_0$ should meet

$$\bar{u}_0 \geq \sqrt{\frac{1}{2}\left(V^2 + \sqrt{V^4 + \left(\frac{VI}{\omega C_d}\right)^2}\right)} \tag{4.43}$$

Finally, the constraint of the average voltage is obtained as

$$\sqrt{\frac{1}{2}\left(V^2 + \sqrt{V^4 + \left(\frac{VI}{\omega C_d}\right)^2}\right)} \leq \bar{u}_0 \leq \sqrt{V_M^2 - \frac{VI}{2\omega C_d}} \tag{4.44}$$

As for the switch-multiplexing circuit topology with single decoupling capacitors, in addition to the above two constraints, there is another constraint, that is, the synthesis of input current and decoupling current should be accomplished in each switching period ($\sum d_j$ should be no more than unity). Hence, inequalities (4.45) should hold for any time.

$$m\left(|\cos(\omega t + \varphi)| + \left|\frac{V\cos(2\omega t + \varphi)/2}{\sqrt{\bar{u}_0^2 + \frac{VI \sin(2\omega t + \varphi)}{2\omega C_d}}}\right|\right) \leq 1 \tag{4.45}$$

where, $m = I/i_{dc}$ is the modulation coefficient.

Then, another constraint is expressed as

$$\sqrt{\left(\frac{Vm}{2(1-m)}\right)^2 + \frac{VI}{2\omega C_d}} \leq \bar{u}_0 \tag{4.46}$$

Thus, $\bar{u}_0$ should meet

$$\begin{cases} \max(a, b) \leq \bar{u}_0 \leq \sqrt{V_M^2 - \frac{VI}{2\omega C_d}} \\ a = \sqrt{\frac{1}{2}\left(V^2 + \sqrt{V^4 + \left(\frac{VI}{\omega C_d}\right)^2}\right)} \\ b = \sqrt{\left(\frac{Vm}{2(1-m)}\right)^2 + \frac{VI}{2\omega C_d}} \end{cases} \tag{4.47}$$

It can be found that the average decoupling capacitor voltage is affected by the modulation coefficient $m$ in the decoupling circuit with single decoupling capacitor. However, this doesn't exist in the decoupling circuit with two decoupling capacitors.

**Table 4.3** Parameters used in analysis, simulation, and experiment

| Parameters | Symbols | Values |
|---|---|---|
| Input phase voltage | $V$ | $110\sqrt{2}$ V |
| Source angular frequency | $\omega$ | 314 rad/s |
| Input filters | $L_i/C_i$ | 0.6 mH/20 $\mu$F |
| Dc filter inductor | $L_{dc}$ | 5 mH |
| Active buffer capacitor | $C_d$ | 90 $\mu$F |
| Load resistance | R | 8.7 $\Omega$ |
| Switching frequency | $f_s$ | 20 kHz |

**Fig. 4.12** Relationship between voltage stress of decoupling capacitor and modulation coefficient $m$ and decoupling capacity $C_d$

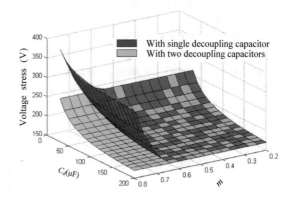

With the same parameters listed in Table 4.3, the voltage stress of the decoupling capacitance is shown in Fig. 4.12. As seen, the voltage stress under the double capacitors case is always less than that under the single capacitor case. For the double capacitors case, the voltage stress is only related to the decoupling capacity. For the single capacitor case, when the modulation coefficient $m$ is less than 0.66 (a > b in (4.47)), the voltage stress is independent of the modulation coefficient. When $m$ is greater than 0.66 (a < b in (4.47)), the voltage stress is related to $m$.

In conclusion, when the modulation coefficient is small, the decoupling circuit with a single capacitor is more suitable as a result of reducing the usage of the hardware and power losses. When the modulation coefficient is relatively large, the decoupling circuit with double capacitors is a good choice due to reducing the voltage stress.

## 4.4    Simulation and Experimental Results

In order to verify the correctness of the theoretical analysis, simulations under ideal conditions and physical experiments are carried out. And the used circuit parameters are listed in Table 4.3.

### 4.4.1    Double Capacitors Case

Figure 4.13 shows the steady-state simulation results under unity power factor operation. The DC-link current reference is set to 7 A. The decoupling capacitor rated at 90 μF is employed and $\bar{u}_0$ is selected to 230 V with a proper safe margin. As can be observed, the DC-link current $i_{dc}$ is flat with only very small fluctuations with the proposed decoupling solution. The input current $i_g$ is sinusoidal and in phase with the input voltage $u_g$. The waveforms of $u_1$ and $u_2$ are in accordance with theoretical analysis as shown in Fig. 4.5 exactly.

As for the experiments the control algorithm of the converter is realized by a combination of digital signal processor TMS320F28335 and field programmable gate array FPGA EP2C8T144C8N. Each decoupling capacitor is formed by connecting three 30 μF/490 V film capacitors in parallel. The experimental results in steady state and transient state are presented.

Figure 4.14a shows the steady-state experimental waveforms. Clearly, the DC-link current $i_{dc}$ is almost constant with only a little high frequency harmonics because of the decoupling function of the capacitors $C_1$ and $C_2$. Meanwhile the low DC-link current ripple verifies the effectiveness of the designed switching patterns. It can also be found that

**Fig. 4.13**  Simulation results

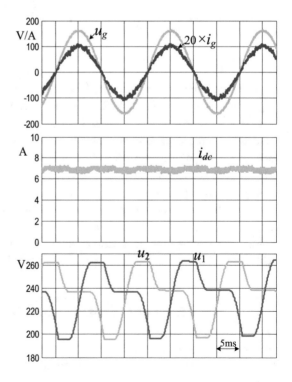

the source current $i_g$ is sine shaped and in phase with the source voltage $u_g$. The THD value of the source current $i_g$ is 4.47% and its harmonic spectrum is given in Fig. 4.14b.

Figure 4.14 shows the waveforms of the decoupling capacitor voltages $u_1$ and $u_2$. Each of them can be divided into six subintervals which is same as that shown in Fig. 4.5. In ideal cases the decoupling capacitor voltages should be fixed when decoupling capacitors are bypassed. However, in practice they drop slightly. The main reasons for the voltage drops include the power losses due to snubber circuits and the self consumption caused by the insulation resistance of the decoupling capacitor. Therefore, as can be seen the capacitor voltages $u_1$ and $u_2$ are not kept constant when capacitors $C_1$ and $C_2$ are bypassed. Furthermore, it is also founded that $u_1$ and $u_2$ are the same in shapes, which is in good agreement with theoretical analysis. Note that the spikes in Fig. 4.15 are introduced in measurements. Because the self-inductance of the ground loop of the used oscilloprobe

**Fig. 4.14** Experimental waveforms of input voltage $u_g$, input current $i_g$, and DC-link current $i_{dc}$ (**a**) and spectral analysis of input current $i_g$ (**b**)

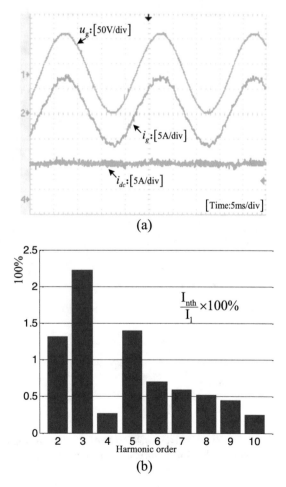

(a)

(b)

is too large, the measured results are affected by the electromagnetic interference caused by the switching transients in measurements.

Figure 4.16 shows the experimental results when the input current leads and lags the input voltage $\pi/18$. As shown in Fig. 4.16a, the duration of Sub-I is longer than that of Sub-II in energy released mode when $\varphi = \pi/18$. Whereas as shown in Fig. 4.16b, the contrary is the case when $\varphi = -\pi/18$. According to previous analysis results in section III, regarding to $\varphi = \pi/18(-\pi/18)$ the durations of Sub-I(Sub-II) and Sub-II(Sub-I) are $5\pi/(18\omega)$ and $2\pi/(9\omega)$, respectively. The experimental results are accordance with the theoretical analysis. In both cases the ripple power is well buffered by the capacitors $C_1$ and $C_2$ and the DC-link current $i_{dc}$ is well kept constant.

**Fig. 4.15** Experimental waveforms of decoupling capacitor voltages $u_1$ and $u_2$

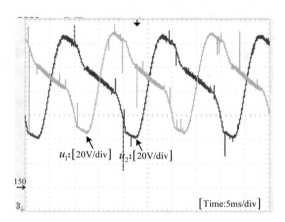

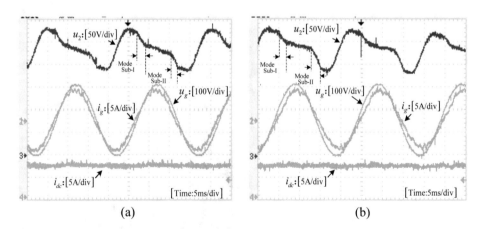

(a)                                              (b)

**Fig. 4.16** Experimental waveforms of decoupling capacitor voltage $u_2$, input voltage $u_g$, input current $i_g$, and DC-link current $i_{dc}$ in the proposed converter. **a** $\varphi = \pi/18$. **b** $\varphi = -\pi/18$

Figure 4.17 illustrates the experimental waveforms when the decoupling function is disabled abruptly. As can been seen, the DC-link current $i_{dc}$ has a large fluctuation immediately, which the peak-peak value is up to 9 A and corresponds to 128.6% of the nominal DC-link current because the ripple power is buffered by the DC-link inductor. Moreover, the spectral analysis of the DC-link current $i_{dc}$ is shown in Fig. 4.18 for the cases with and without activating the decoupling function. It is obvious that the second-order harmonic component, i.e. 100 Hz, is significantly reduced with activating the decoupling function. To realize the same current ripple level without activating the decoupling function the required inductance is 440 mH. So the proposed circuit improves system power density significantly. In addition, other order harmonic components in the DC-link current with activating the decoupling function are much lower than those without activating the decoupling function.

Figure 4.19 shows the dynamic response of the system when the load power subjects from 50 to 100% step-up change. As can be seen, the DC-link current $i_{dc}$ tracked the reference well and quickly with the proposed control algorithm. And the transient process is smooth and there is no obvious distortion in the input current. The fluctuation ranges of $u_1$ and $u_2$ increase accordingly due to the increased ripple power.

Figure 4.20 illustrates the efficiency curves under different load power with/without the decoupling circuit. The tests have been done under unity power factor. As can be seen, the efficiency of the proposed converter is lower than that of the traditional SCSC. This is because the increased equivalent switching frequency (increased by 1/2) and voltage stresses contribute to additional switching power losses, and the conduction power losses are also increased.

**Fig. 4.17** Dynamic experimental waveforms by disabling the decoupling function abruptly

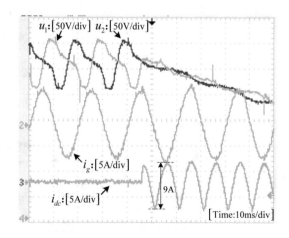

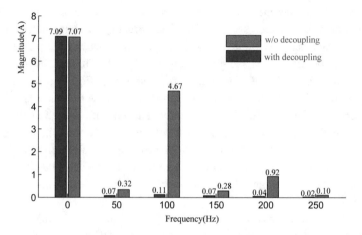

**Fig. 4.18** Spectral analysis of the DC-link current with/without decoupling

**Fig. 4.19** Dynamic
experimental waveforms
showing 50–100% step-up load
change

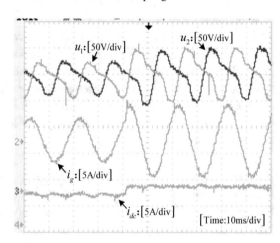

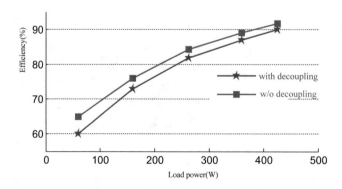

**Fig. 4.20** System efficiency curve with/without decoupling

**Fig. 4.21**  Simulation results

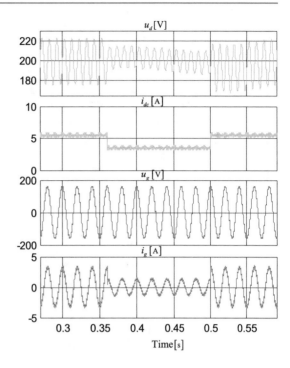

## 4.4.2  Single Capacitor Case

Figure 4.21 shows the simulation waveforms of the decoupling capacitor voltage, the DC-link current, the grid voltage, and the grid current. At the beginning, the given power reference is 250 W and the DC-link current is about 5.4 A. The step-down change from 250 to 100 W and the step-up change 100 W to 250 W are carried out at $t = 0.36$ s and 0.5 s, respectively. As seen, the DC-link current $i_{dc}$ is always smooth because the ripple power is diverted to the decoupling capacitor $C_d$. Meanwhile, the low DC-link current ripple verifies the effectiveness of the designed switching patterns. The source current $i_g$ is always sine shaped and in phase with the source voltage $u_g$.

Figure 4.22a shows the steady-state experimental waveforms. It can be found that the experimental results are in good agreement with the simulation results. The THD of source current $i_g$ is 4.63%. After disabling the decoupling function, the twice ripple power will be imposed on the DC-link inductor immediately and the related experimental waveforms are shown in Fig. 4.22b. Clearly, the DC-link current fluctuates with twice the line frequency. And the grid ac current is seriously distorted at the valley of the DC-link current. Figure 4.22c shows the spectrum of the DC-link current with and without decoupling. It is clear that there is a dramatic reduction of the 2nd harmonic current when activating decoupling function.

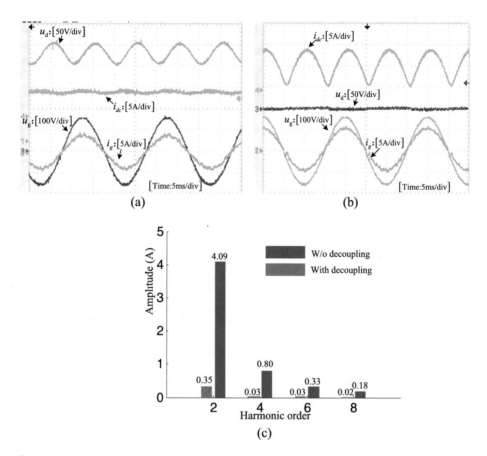

**Fig. 4.22** Steady-state experimental waveforms. **a** With decoupling function. **b** Without decoupling function. **c** Spectral analysis for DC-link current

Figure 4.23 shows the waveforms of voltages across $(D_4 + S_4)$ and $D_2$. As seen, the envelope of the voltage across $(D_4 + S_4)$ is $(u_g + u_d)$, which is always positive. Therefore, the reverse voltage across $D_4$ is negative during the turn-off process, which reduces the turn-off loss. The envelope of the voltage across $D_2$ is $(-u_g - u_d)$, which is always negative. Then $D_2$ can be blocked reliably when $S_4$ is in the on state. On the other hand, the voltage stress of $S_4$ and $D_2$ are the highest in all the semiconductor devices.

Figure 4.24 shows the dynamic response of the system when the load power is subject to a 100–40% step-down change and a 40–100% step-up change. As can be seen, in Fig. 4.24a the DC-link current $i_{dc}$ decreases from 5.4 to 3.4 A quickly due to the closed-loop control method. The excess energy in the DC-link inductor is transferred to the decoupling capacitor and the capacitor voltage level is raised. The decoupling capacitor voltage enters steady state relatively slow as results of low bandwidth of the voltage

**Fig. 4.23** Experimental
waveforms of voltages across
($D_4 + S_4$) and $D_2$

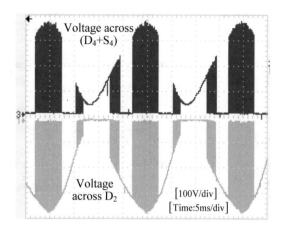

control loop. With the decrease of the load power, the ripple power is reduced and the
fluctuation range of $u_d$ decreases accordingly. Figure 4.24b shows the opposite transient
process. The experimental results also coincide with those in the simulation.

Figure 4.25a illustrates the efficiency curves of the proposed converter and the conventional single-phase current source rectifier (SCSR). As can be seen, the efficiency of
the proposed converter is slightly lower than that of the conventional case. The main reason is the increased voltage stresses, which increases the switching power losses. As well
known, all the active decoupling methods cause extra power losses. Usually, the efficiency
penalty is more than two percentage points. However, in the proposed method the system
efficiency drop under rated load power is 1.5%. Figure 4.25b shows the estimated power
loss distribution. Apparently, the switching losses of the semiconductor devices in the

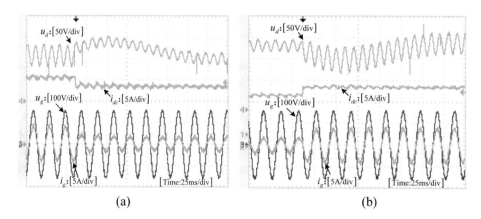

**Fig. 4.24** Dynamic experimental waveforms. **a** 100–40% step-down load change. **b** 40–100% step-
up load change

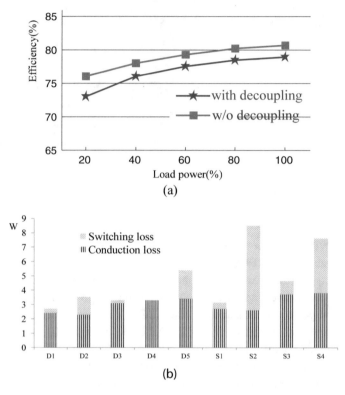

**Fig. 4.25** Power losses analysis. **a** Efficiency comparison. **b** Loss distribution of the proposed converter with 100% load operation

lower bridge arm are significant. As analyzed previously, switching loss doesn't happen to $D_4$ because the reverse voltage across $D_4$ is negative.

## 4.5    Other Dependent Decoupling Topologies

Figure 4.26 shows some other dependent decoupling topologies. Figure 4.26a shows a switch-multiplexing decoupling circuit formed by sharing two lower switches between the decoupling unity and the rectifier. It is termed as the vertical multiplexing decoupling topology. Its main advantage is the large conduction power losses due to fact that six semiconductor devices are involved in the current path.

Figure 4.26b–d show switch-multiplexing decoupling circuits formed by sharing one bridge-arm between the decoupling unity and the rectifier. It is termed as the horizontal multiplexing dependent decoupling topology. In Fig. 4.26b the decoupling capacitor voltage $u_d$ can be positive or negative. If it is limited to only withstand the DC voltage,

**Fig. 4.26** Other dependent decoupling topologies. **a** Vertical multiplexing dependent decoupling topology. **b** Horizontal multiplexing dependent decoupling topology (version I). **c** Horizontal multiplexing dependent decoupling topology (version II). **d** Horizontal multiplexing dependent decoupling topology (version III)

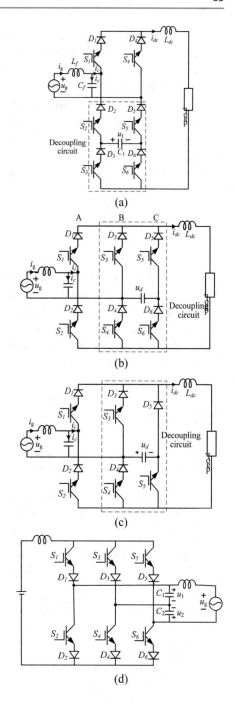

the converter shown in Fig. 4.26c can be obtained by removing redundant switches and diodes. Note that in Fig. 4.26c the decoupling capacitor voltage needs to be higher than the peak grid voltage to avoid the misgating-on of the diodes. The horizontal multiplexing decoupling circuits reduce the conduction power losses as a result of less switches included in the current path. However, the DC-link current utilization ratio is reduced.

With adopting the differential connection operation applied for the decoupling capacitor and the filter capacitor, another horizontal multiplexing version is obtained as shown in Fig. 4.26d. The differential voltage ($v_1$–$v_2$) is equal to the grid voltage in steady state. And the common voltage is used to buffered the ripple power, which results in low volage stress.

## 4.6    Conclusion

This chapter introduces two kinds of switching-multiplexed decoupling circuits, which require no additional switching devices. Compared with the independent decoupling topology introduced in Chap. 3, they need less hardware. The decoupling with a single capacitor is suitable for small modulation coefficient case, and the decoupling circuit with double capacitors is more suitable for large modulation coefficient case. Some other dependent decoupling topologies are also introduced in this chapter.

## Bibliography

1. Cai, W., Jiang, L., Liu, B., et al. (2015). A power decoupling method based on four-switch three-port DC-DC-AC converter in DC microgrid. *IEEE Transactions on Industry Applications, 51*(1), 336–343.
2. Han, H., Liu, Y., Sun, Y., Su, M., & Xiong, W. (2015). Single-phase current source converter with power decoupling capability using a series-connected active buffer. *IET Power Electronics, 8*(5), 700–707.
3. Vitorino, M. A., & de Rossiter Correa, M. B. (2014). Compensation of DC-link oscillation in single-phase VSI and CSI converters for photovoltaic grid connection. *IEEE Transactions on Industry Applications, 50*(3), 2021–2028.
4. Vitorino, M. A., Hartmann, L. V., Fernandes, D. A., Silva, L. E., & Correa, M. B. R. (2014). Single-phase current source converter with new modulation approach and power decoupling. In *Proceedings of the IEEE APEC*, Fort Worth, TX 3:2200–2207.
5. Vitorino, M. A., Correa, M. B. R., & Jacobina, C. B. (2013). Single-phase power compensation in a current source converter. In *Proceedings of the IEEE ECCE*, Denver, CO (pp. 5288–5293).
6. Qi, W., Wang, H., Tan, X., et al. (2014). A novel active power decoupling single-phase PWM rectifier topology. In *Proceedings of the IEEE APEC*, Fort Worth, TX (pp. 89–95).
7. Serban, I. (2015). Power decoupling method for single-phase H-bridge inverters with no additional power electronics. *IEEE Transactions on Industrial Electronics, 62*(8), 4805–4813.

8. Tang, Y., & Blaabjerg, F. (2015). A component-minimized single-phase active power decoupling circuit with reduced current stress to semiconductor switches. *IEEE Transactions on Power Electronics, 30*(6), 2905–2910.

9. Sun, Y., Liu, Y., Su, M., Li, X., & Yang, J. (2016). Active power decoupling method for single-phase current-source rectifier with no additional active switches. *IEEE Transactions on Power Electronics, 31*(8), 5644–5654.

10. Sun, Y., Liu, Y., Su, M., Li, X., & Yang, J. (2016). Active power decoupling method for single-phase current source rectifier with no additional active switches. *IEEE Transactions on Power Electronics, 31*(8), 5644–5654.

11. Liu, Y., Sun, Y., & Su, M. (2016). Active power compensation method for single-phase current source rectifier without extra active switches. *IET Power Electronics, 9*, 1719–1726.

12. Zhu, G. R., Tan, S. C., Chen, Y., et al. (2013). Mitigation of low-frequency current ripple in fuel-cell inverter systems through waveform control. *IEEE Transactions on Power Electronics, 28*(2), 779–792.

# Decoupling Topologies with Wide Output Voltage

**5**

The original intent of active power decoupling circuit is only to buffer the ripple power and therefore improve system reliability and power density. As research continues, more and more functions have been merged into the decoupling circuits. Then, the converter achieves multi-functions with less semiconductor devices and passive components. That is beneficial for saving cost. As well known, SCSR is a buck-type rectifier in essence. And the DC output voltage is limited to half the peak grid voltage. If a higher output voltage is required, a rear boost converter is needed. That will increase cost and power losses inevitably. This chapter presents a family of two-port switching networks (TSNs) to decouple the ripple power and boost the output voltage simultaneously. Section 5.1 introduces the output voltage range of the SCSR. And a current-source rectifier with decoupling and voltage boost function is presented in Sect. 5.2. In Sect. 5.3, its switching states and operation principle are analyzed. Then, the selection law of the decoupling capacitor is explained in Sect. 5.3. Section 5.4 gives simulation and experimental results to show the effectiveness. A single-phase inverter with wide input voltage and power decoupling capability is introduced and discussed in Sect. 5.5.

## 5.1 Output Voltage Range of SCSR

According to the law of power conservation, we have

$$u_L i_{dc} = V I \cos(\varphi)/2 \tag{5.1}$$

where $u_L$ is the output voltage, $i_{dc}$ is the DC bus current, $V$ and $I$ are respectively the amplitudes of the gird voltage and current, and $\varphi$ is the displacement angle. Then the voltage transmission ratio $\lambda$ can be expressed as

© The Author(s), under exclusive license to Springer Nature Switzerland AG 2023    87
Y. Liu, *Active Power Decoupling Technology in Single-Phase Current-Source Converters*,
Synthesis Lectures on Power Electronics, https://doi.org/10.1007/978-3-031-21270-3_5

**Fig. 5.1** SCSR with a boost circuit to achieve a wider output voltage

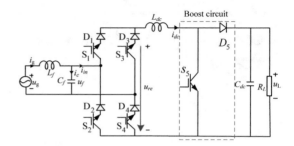

$$\lambda = \frac{u_L}{V} = \frac{m \cos(\varphi)}{2} \tag{5.2}$$

where $m = I/i_{dc}$ ($0 \leq m \leq 1$) is the input current modulation coefficient of the SCSR. Obviously, $\lambda \leq 0.5$. Hence, the SCSR is only suitable for low voltage applications. If a large output voltage is required, an additional DC-DC boost circuit has to be employed as shown in Fig. 5.1. Besides, the DC-link inductor $L_{dc}$ should be large to smooth the DC-link current $i_{dc}$.

## 5.2    SCSR with Decoupling and Voltage Boost Function

The investigated SCSR with integrating decoupling and voltage boost function is shown in Fig. 5.2a. It is formed by inserting a TSN into the DC-link of a conventional SCSR. The TSN plays dual roles of buffering the ripple power and boosting the output voltage. Four different TSN topologies are presented, as shown in Fig. 5.2b–e. All of them consist of two switches $S_5$ and $S_6$, two diodes $D_5$ and $D_6$, and a ripple energy buffering capacitor $C$. They are divided into type I and type II. Each type includes two kinds of TSNs, which are identical in essence.

To reduce switching losses, the active switches $S_5$ and $S_6$ can be replaced with MOS-FETs. On the other hand, if only unity power factor is required, the front-end active rectifier can be replaced with a diode rectifier with PFC.

## 5.3    Switching States and Operation Principle

### 5.3.1    Switching States

As the two kinds of TSNs in each type are identical, only the operation modes of type I-1 and II-1 are analyzed, as shown in Figs. 5.3 and 5.4, respectively. The front rectifier is equivalent to a controlled voltage source $u_{re}$ and the solid line indicates the DC-link current pathways.

**Fig. 5.2** Proposed SCSR.
**a** General circuit structure with
a TSN. **b** Type I-1. **c** Type I-2.
**d** Type II-1. **e** Type II-2

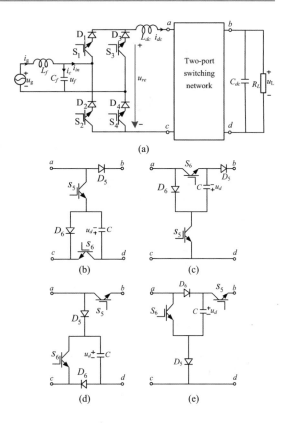

Figure 5.3 shows the four operation modes of Type I-1. Modes I and II are designed for free-wheeling purpose. In both of them the switch $S_5$ is turned on and the load is bypassed. In mode II, the decoupling capacitor $C$ is discharged. Modes III and IV are designed for the power transmission purpose. In both of them switch $S_5$ is turned off and the DC-link current flows through the load. In mode IV, the decoupling capacitor is charged. As can be seen, $S_5$ determines whether the DC-link current flows through the load and $S_6$ determines whether the DC-link current flows through the decoupling capacitor.

Figure 5.4 shows the four operation modes of type II-1. Differently, modes I and II are designed for the power transmission, while modes III and IV are designed for the free-wheeling. However, the decoupling capacitor voltage $u_d$ should be larger than the output voltage to ensure that $D_5$ is reverse biased in modes I and II and avoid damaging $S_5$ in modes III and IV. Then the decoupling capacitor voltage has a higher voltage level compared with that in type I-1, which causes higher voltage stresses on $S_6$ and $D_6$. On the other hand, the voltage stresses of $S_5$ and $D_5$ are $(u_d - u_L)$ which may be smaller than those $(u_d + u_L)$, in type I-1.

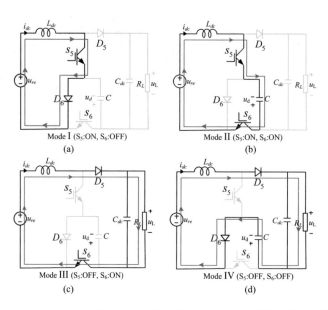

**Fig. 5.3** Operation modes of type I-1. **a** Mode I. **b** Mode II. **c** Mode III. **d** Mode IV

**Fig. 5.4** Operation modes of type II-1. **a** Mode I. **b** Mode II. **c** Mode III. **d** Mode IV

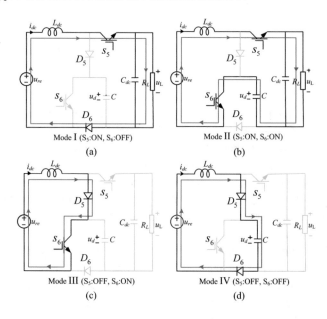

One distinct difference between the two types of TSNs lies in the decoupling capacitor voltage level. In type I the decoupling capacitor voltage is free from the output voltage; while in type II it has to be larger than the output voltage. As the operation principles of the four TSNs are similar, only type I-1 is taken as an example to be analyzed in detail in the following sections.

### 5.3.2   Operation Principle

The ac equivalent circuit of the type I-1 converter is shown in Fig. 5.5. $d_i$ ($i = 1, 2, 3, 4$) is the duty ratio of mode $i$ and $d_1 + d_2 + d_3 + d_4 = 1$. $u_{ab} = d_d u_d$ is an averaged voltage provided by decoupling capacitor. $D$ and $d_d$ are expressed as

$$D = d_3 + d_4 \tag{5.3}$$

$$d_d = d_4 - d_2 \tag{5.4}$$

where $0 < D \leq 1$.

Assume that the source ac voltage $u_g$ is $V \cos(\omega t)$ and ac current $i_g$ is $I \cos(\omega t + \varphi)$. Then the instantaneous input power $P$ can be expressed as $u_g i_g$. Suppose the power processing by TSN is $P_d$ and the load power is $P_L$, then according to power balance we have

$$P = P_d + P_L = \frac{VI \cos(2\omega t + \varphi)}{2} + \frac{VI \cos(\varphi)}{2} \tag{5.5}$$

$P_d$ and $P_L$ can be further expressed as

$$P_d = d_d i_{dc} u_d = \frac{VI \cos(2\omega t + \varphi)}{2} \tag{5.6}$$

$$P_L = D i_{dc} u_L = \frac{VI \cos(\varphi)}{2} \tag{5.7}$$

**Fig. 5.5** Equivalent circuit of the type I-1 converter

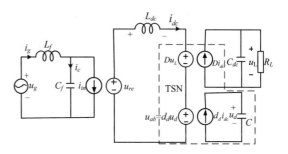

In (5.6), $d_d$ is a variable which can be positive or negative. When $d_d$ is greater than zero, the decoupling capacitor is charged (absorbing the second-order ripple power). Otherwise, the decoupling capacitor is discharged (releasing the second-order ripple power).

According to (5.7), the output voltage $u_L$ can be obtained as

$$u_L = \frac{VI\cos(\varphi)}{2Di_{dc}} \tag{5.8}$$

Suppose $i_{dc}$ is controlled to be a given constant, $D$ should be constant to achieve a constant output voltage. Then the voltage transfer ratio $\lambda$ in type I-1 is

$$\lambda_1 = \frac{m\cos(\varphi)}{2D} \tag{5.9}$$

According to (5.2) and (5.9), $\lambda_1$ is always larger than $\lambda$ under the same $m$ because the proposed converter provides another degree of freedom $D$ to regulate the output voltage. That is the reason why it could output a wider range voltage.

The dynamic differential equation of the decoupling capacitor is

$$C\frac{du_d}{dt}u_d = P_d \tag{5.10}$$

Combining (5.6) with (5.10), the decoupling capacitor voltage is expressed as

$$u_d = \sqrt{u_0^2 + \frac{VI}{2\omega C}\sin(2\omega t+\varphi)} \tag{5.11}$$

where $u_0$ is the DC component.

According to (5.6) and (5.9), it is easy to obtain $d_d$ and $D$. To reduce the switching times, when charging the decoupling capacitor, mode II is not used. While, when discharging the decoupling capacitor, mode IV is not used. Then, referring to (5.3) and (5.4), $d_i$ is given as

$$\begin{cases} d_1 = 1 - \sum\limits_{i=2}^{4} d_i, & d_2 = \begin{cases} 0, & d_d > 0 \\ -d_d, & d_d \leq 0 \end{cases} \\ d_3 = D - d_4, & d_4 = \begin{cases} d_d, & d_d > 0 \\ 0, & d_d \leq 0 \end{cases} \end{cases} \tag{5.12}$$

According to (5.12), the applied switching patterns in TSN are shown in Fig. 5.6, where $T_s$ is the switching period.

**Fig. 5.6** Switching patterns. **a** Charging C. **b** Discharging C

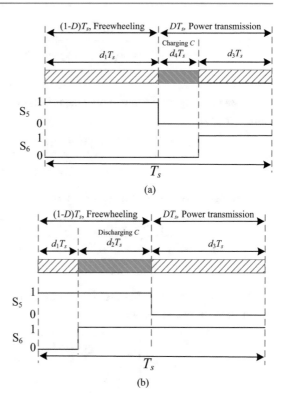

(a)

(b)

## 5.4 Analysis of the Decoupling Capacitor Voltage

As can be seen from Fig. 5.6, charging duration $d_4 T_s$ should be less than $DT_s$ and discharging duration $d_2 T_s$ should be less than $(1 - D)T_s$. Then the following inequations can be obtained

$$0 \le d_2 \le 1 - D \tag{5.13}$$

$$0 \le d_4 \le D \tag{5.14}$$

According to (5.12), duty cycles $d_2$ and $d_4$ can be rewritten as

$$d_2 = \begin{cases} \dfrac{-D\lambda_1 V \cos(2\omega t + \varphi)}{\cos(\varphi)\sqrt{u_0^2 + \frac{VI}{2\omega C} \sin(2\omega t + \varphi)}}, & \cos(2\omega t + \varphi) < 0 \\ 0, & \cos(2\omega t + \varphi) \ge 0 \end{cases} \tag{5.15}$$

$$d_4 = \begin{cases} \dfrac{D\lambda_1 V \cos(2\omega t + \varphi)}{\cos(\varphi)\sqrt{u_0^2 + \frac{VI}{2\omega C}\sin(2\omega t + \varphi)}}, & \cos(2\omega t + \varphi) \geq 0 \\ 0, & \cos(2\omega t + \varphi) < 0 \end{cases} \tag{5.16}$$

When discharging the decoupling capacitor, i.e., $\cos(2\omega t + \varphi) < 0$, according to (5.13), one have

$$\frac{-D\lambda_1 V \cos(2\omega t + \varphi)}{\cos(\varphi)\sqrt{u_0^2 + \frac{VI}{2\omega C}\sin(2\omega t + \varphi)}} \leq 1 - D \tag{5.17}$$

Then one lower limit of $u_0$ can be obtained

$$u_{01} = \begin{cases} \dfrac{1}{2(1-D)}\sqrt{\dfrac{\cos^2(\varphi)(1-D)^4 I^2}{(2\omega C D\lambda_1)^2} + \dfrac{(2VD\lambda_1)^2}{\cos^2(\varphi)}}, & \dfrac{I\cos^2(\varphi)(1-D)^2}{4\omega C V D^2 \lambda_1^2} \leq 1 \\ \sqrt{\dfrac{VI}{2\omega C}}, & \dfrac{I\cos^2(\varphi)(1-D)^2}{4\omega C V D^2 \lambda_1^2} > 1 \end{cases} \tag{5.18}$$

Similarly, when charging the decoupling capacitor, i.e., $\cos(2\omega t + \varphi) \geq 0$, the other lower limit of $u_0$ can be obtained

$$u_{02} = \begin{cases} \dfrac{1}{2D}\sqrt{\dfrac{\cos^2(\varphi)D^4 I^2}{(2\omega C D\lambda_1)^2} + \dfrac{(2VD\lambda_1)^2}{\cos^2(\varphi)}}, & \dfrac{I\cos^2(\varphi)D^2}{4\omega C V D^2 \lambda_1^2} \leq 1 \\ \sqrt{\dfrac{VI}{2\omega C}}, & \dfrac{I\cos^2(\varphi)D^2}{4\omega C V D^2 \lambda_1^2} > 1 \end{cases} \tag{5.19}$$

Suppose the maximum admissible decoupling capacitor voltage is $u_p$, which is determined by the voltage stresses of the selected decoupling capacitor and switches. Then the upper limit of $u_0$ is

$$u_0 \leq \sqrt{u_p^2 - \frac{VI}{2\omega C}} \tag{5.20}$$

Then the admissible range of the average voltage across the decoupling capacitor $u_o$ can be expressed as

$$\max\{u_{01}, u_{02}\} \leq u_0 \leq \sqrt{u_p^2 - \frac{VI}{2\omega C}} \tag{5.21}$$

As for type II converters, another limitation is that the decoupling capacitor voltage $u_d$ should be greater than the output voltage $u_L$. Therefore, the admissible range of the average voltage $u_0$ is

$$\max\{u_{01}, u_{02}, u_{03}\} \leq u_0 \leq \sqrt{u_p^2 - \frac{VI}{2\omega C}} \tag{5.22}$$

where $u_{03} = \sqrt{\lambda^2 V^2 + \frac{VI}{2\omega C}}$.

Based on (5.11), (5.21), and (5.22), the decoupling capacitor voltage stresses (normalized by $V$) with respect to $C$ and $D$ are shown in Fig. 5.7 using the parameters in Table 5.1. As seen, it is better for the system to operate around $D = 0.5$ for the view of decreasing the voltage stress. In addition, it can be found that when the decoupling capacitance is increased to a certain degree, increasing capacitance cannot decrease the voltage stress dramatically. By making a trade-off between voltage stress and cost, the decoupling capacitance is selected to be 100 μF.

On the other hand, for type II $u_{03}$ dominates the lower limit around $D = 0.5$. Then, the average voltage across the decoupling capacitor is greater than that in type I. While, $u_{01}$

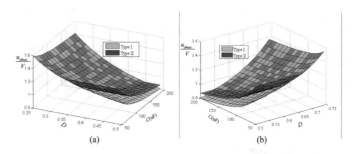

**Fig. 5.7**  Decoupling capacitor voltage stress as a function of the decoupling capacitance $C$ and duty ratio $D$. **a** $D$ varies from 0.25 to 0.5. **b** $D$ varies from 0.5 to 0.75

**Table 5.1**  Parameters used in analysis, simulation, and experiment

| Parameters | Symbols | Values |
|---|---|---|
| Output power | $P_L$ | 280 W |
| Input phase voltage | $V$ | 110 Vrms |
| Source angular frequency | $\omega$ | 314 rad/s |
| Input filters | $L_f/C_f$ | 0.6 mH/20 μF |
| DC-link filter inductor | $L_{dc}$ | 3 mH |
| Output filter capacitor | $C_{dc}$ | 10 μF |
| Active buffer capacitor | $C$ | 100 μF |
| Load | $R$ | 40 Ω |
| Switching frequency | $f_s$ | 20 kHz |

($u_{02}$) dominates the lower limit with increasing (decreasing) $D$. Then, the average voltage is equal to that in type I.

## 5.5     Simulation and Experimental Results

### 5.5.1   Simulation Results

Simulation study is carried out using Piecewise Linear Electrical Circuit Simulation (PLECS). The main parameters are listed in Table 5.1. $u_0$ is selected to 130 V and 160 V in type I-1 and type II-1, respectively.

   Figures 5.8 and 5.9 show the steady-state operation waveforms of type I-1 and type II-1 circuit topologies. From the top, the grid voltage $u_g$, the grid current $i_g$, output voltage $u_L$, output current $i_L$, decoupling capacitor voltage $u_d$, and DC-link current $i_{dc}$ are sequentially displayed. As can be observed, in both figures the output voltage $u_L$ is about 105 V which is larger than half peak amplitude of the ac source voltage (75.5 V). The source current $i_g$ is always sine shaped and in phase with the source voltage $u_g$. The DC-link current $i_{dc}$ is approximately constant. That indicates the effectiveness of the used closed-loop control method. The only difference is that in Fig. 5.9 the decoupling capacitor voltage $u_d$ is always greater than the output voltage to ensure normal operation, while in Fig. 5.8 it can be smaller or larger than the output voltage. So the voltage stress of the decoupling capacitor in type II-1 is higher than that in type I-1.

### 5.5.2   Experimental Results

Figure 5.10 shows the experimental waveforms of type I-1 circuit topology. Figure 5.10a, b are steady state waveforms of grid voltage/current, output voltage/current, decoupling capacitor voltage, collector-emitter voltage of switch $S_5$, and DC-link current. As can be seen, the experimental waveforms are in good agreement with the simulation results in Fig. 5.8. The envelope of the voltage across $S_5$ is ($u_d + u_L$), which is in consistence with the theoretical analysis. When the decoupling function is suddenly disabled ($S_6$ is always turned on and $S_5$ is modulated by a pulse waveform with the duty ratio $D$), the TSN works as a boost circuit and the transient waveforms are shown in Fig. 5.10c. As seen, the energy stored in the decoupling capacitor is completely consumed by the insulation resistance and the decoupling capacitor voltage decreases to zero. The DC-link current $i_{dc}$ has a large fluctuation immediately, and the peak-peak value is up to 11.5 A due to the fact that the DC-link inductor buffers a part of ripple power. Moreover, the grid current has significant harmonics.

   Figure 5.11 shows the experimental waveforms of type II-1 circuit topology. The steady state experimental waveforms shown in Fig. 5.11a, b are in good agreement with the

**Fig. 5.8** Steady-state
simulation results of type I-1

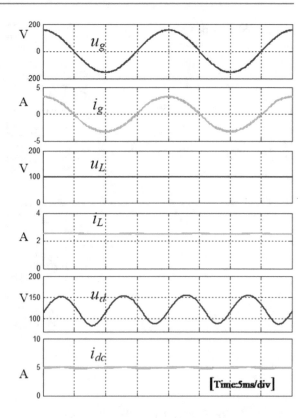

simulation results in Fig. 5.9. From Fig. 5.11b the envelope of the voltage across $S_5$ is ($u_d$ $- u_L$), which is beneficial for reducing switching losses. Figure 5.11c shows the transient waveforms when deactivating the decoupling function suddenly ($S_6$ is always turned on and $S_5$ is modulated by a pulse waveform with the duty ratio $D$). Similarly, the DC-link current $i_{dc}$ has a large fluctuation of 11.5 A (peak-peak value) and distinct distortion exists in the grid current. The difference from Fig. 5.11c is that the decoupling capacitor voltage won't drop to zero. That is because the energy in the load will be transferred to the decoupling capacitor to keep its voltage being equal to the output voltage when $S_5$ is turned on.

Figure 5.12a shows the spectral analysis of the DC-link current with type I-1 topology, type II-1 topology, and the one disabling decoupling function. It is obvious that the second-order harmonic component, i.e., 100 Hz, is significantly reduced with the proposed circuits. In addition, other order harmonic components in the DC-link current with activating decoupling function are much lower as well. It also can be found that the decoupling performance in type II-1 is a little better than that in type I-1. Figure 5.12b shows the overall efficiency curves. The load power is increased by improving output voltage. Clearly, the proposed converters have lower conversion efficiency than the conventional

**Fig. 5.9**  Steady-state
simulation results of type II-1

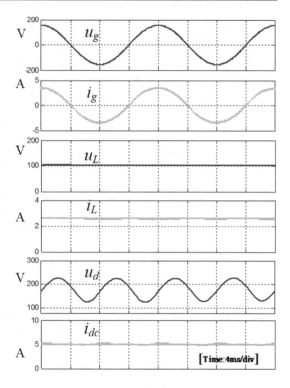

method (SCSR followed by a boost converter). The added power losses are caused by
TSN. On the other hand, the efficiency of type II-1 is little higher than that of type I-1. It
also can be found that the efficiency measured as $D = 0.5$ is higher than that measured
at $D = 0.35$. The reason is that the smaller $D$ raises the voltage stress, which is analyzed
in Fig. 5.7.

Table 5.2 gives a comparison between different SCSRs with power decoupling function. It can be found that the cost of the proposed topologies is comparable to that of
most other SCSRs. And the highlight of the proposed topologies is the wide output voltage range. Besides, the control and modulation designs are flexible because the TSN and
the original rectifier work independently. Therefore, the proposed topologies can be good
candidates for buffering the ripple power, especially when a wide output voltage range is
required.

## 5.6   Single-Phase Inverter with Wide Input Voltage and Power Decoupling Capability

Figure 5.13 outlines single-phase inverter with wide input voltage and power decoupling
capability. Switches $S_1$-$S_4$ switch at twice the grid frequency to change the direction of

**Fig. 5.10** Steady-state experimental results of type I-1. **a** Ac source voltage/current and output voltage/current. **b** Decoupling capacitor voltage, output voltage, switching voltage $u_{ce}$ of $S_5$, and DC-link current. **c** Transient waveforms when disabling the decoupling function

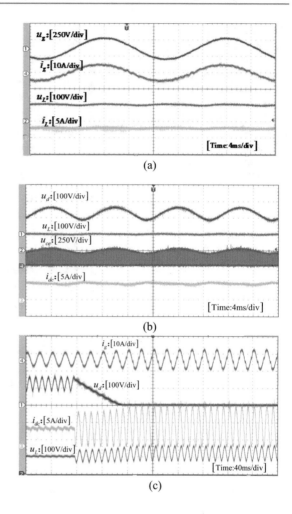

(a)

(b)

(c)

the DC-link current $i_{dc}$ flowing through the filter capacitor $C_a$. $v_r$ is the rectified output voltage and $S_r$ is an equivalent switch, which indicates whether the dc inductor current $i_{dc}$ passes through the grid side. It can be found that when $S_r$ is turned off, the rest circuit is a boost DC-DC circuit from $V_s$ to $V_d$. When $S_r$ is turned on, the diode $D_1$ is reverse-biased and the rest circuit is a buck DC-DC circuit from $V_d$ to $v_r$. Therefore, the whole circuit can be viewed as a buck-boost circuit from $V_s$ to $v_r$. Then, the limit that the dc source voltage is limited to half the grid peak voltage is broken, which extends the scope of applications. At the same time, the inherent low frequency ripple power is buffered by swinging the decoupling capacitor voltage $u_d$.

Figure 5.14 shows the steady-state experimental results. As can be observed, the dc source voltage is 100 V, which is higher than half the peak grid voltage (78 V). The input current $i_g$ is sinusoidal and in phase with the input voltage $v_g$. The PFC is 0.99 and the

**Fig. 5.11** Steady-state experimental results of type II-1. **a** Ac source voltage/current and output voltage/current. **b** Decoupling capacitor voltage, output voltage, switching voltage $u_{ce}$ of $S_5$, and DC-link current. **c** Transient waveforms when disabling the decoupling function

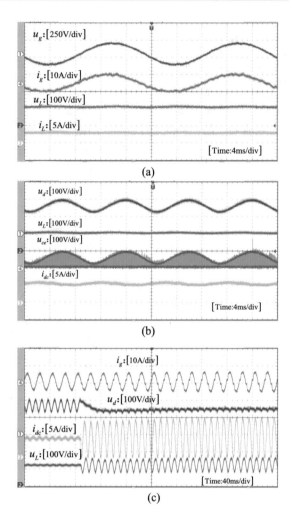

(a)

(b)

(c)

THD is 4.3%. As seen, the dc source current $i_s$ is flat with only very small fluctuations because the ripple power is diverted to the decoupling capacitor $C_d$. Therefore, decoupling capacitor voltage $v_d$ swings at twice the grid frequency and its peak-to-peak is about 100 V.

## 5.7  Conclusion

A family of two-port switching networks is proposed to buffer the second-order ripple power and boost the output voltage at the same time. It breaks the limitation that the maximum output voltage is half the peak amplitude of the ac voltage in the exiting

**Fig. 5.12** Spectral analysis and efficiency. **a** Spectral analysis of the DC-link current with type I-1, type II-1, and the one disabled decoupling function. **b** Efficiency curves of the proposed circuits

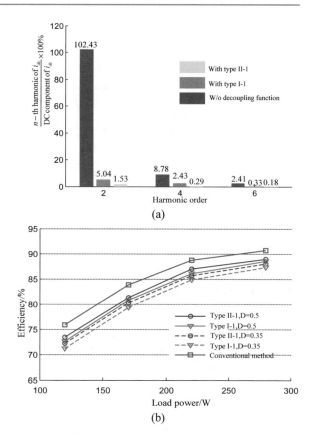

SCSRs. That expands the scope of the proposed circuits, for example, electric vehicles with different voltage levels can be connected to be charged. The performances of the proposed converters are verified by simulations and experimental results. Good decoupling performance indicates the effectiveness of the adopted closed-loop control method.

**Table 5.2** Comparisons between different SCSRs with power decoupling function

| Decoupling technology | Power rating | Decoupling component (value) | Adding semiconductor devices | Efficiency | Maximum DC voltage | Features |
|---|---|---|---|---|---|---|
| Reference [2] | 400 W | 1 Capacitor (50 μF) | 1 Mosfet + 2 Diodes | 94.9% | $V/2$ | – Additional PFC function |
| Reference [3] | 750 W | 1 Capacitor (100 μF) | 2 Mosfets + 1 Diode | 96.4% | $V/2$ | – Decoupling capacitor voltage must be higher than the peak value of grid voltage |
| Reference [4] | 1.5 kW | 1 Capacitor (300 μF) | 2 Switches + 2 Diodes | – | $V/2$ | – Cooperation with original converter<br>– Horizontal multiplexing |
| Reference [5] | 1 kW | 2 Capacitors (32 μF) | 2 IGBTs + 2 Diodes | – | $V/2$ | – Cooperation with original converter<br>– Decoupling capacitors also works as the ac filters |
| Reference [6] | 1 kW | 1 Capacitor (50 μF) | 2 IGBTs + 2 Diodes | – | $V/2$ | – Cooperation with original converter<br>– Vertical multiplexing |
| In this chapter | 280 W | 1 Capacitor (100 μF) | 2 IGBTs + 2 Diodes | See Fig. 5.12b | Unlimited | – Additional buck-boost function<br>– TSN and the original rectifier are independent |

**Fig. 5.13** Proposed inverter (**a**) and its equivalent circuit (**b**)

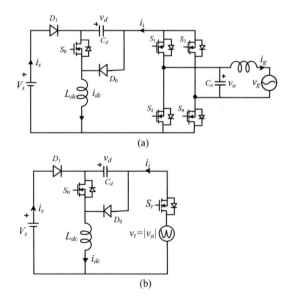

**Fig. 5.14** Steady-state experimental waveforms. **a** Grid voltage $v_g$, grid current $i_g$, dc source voltage $V_S$, and dc source current $i_S$. **b** Grid voltage $v_g$, grid current $i_g$, decoupling capacitor voltage $v_d$, and inductor currents $i_{dc}$

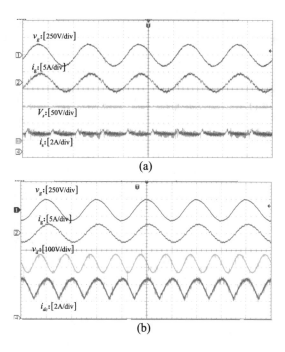

(a)

(b)

# Bibliography

1. Klumpner, C., & Blaabjerg, F. (2006). Using reverse blocking IGBTs in power converters for adjustable speed drives. *IEEE Transactions on Industry Applications, 42*(3), 807–816.
2. Ohnuma, Y., & Itoh, J. I. (2014). A novel single-phase buck PFC AC–DC converter with power decoupling capability using an active buffer. *IEEE Transactions on Industry Applications, 50*(3), 1905–1914.
3. Ohnuma, Y., Orikawa, K., & Itoh, J. I. (2015). A single-phase current-source PV inverter with power decoupling capability using an active buffer. *IEEE Transactions on Industry Applications, 51*(1), 531–538.
4. Saisho, M., Harimoto, T., Hayashi, H., & Saito, M. (2013). Development of single-phase current source inverter with power decoupling function. In *Proceedings of the IEEE International Conference on Power Electronics and Drive Systems*, Kitakyushu, Japan (pp. 591–596).
5. Vitorino, M. A., Correa, M. B. R., & Jacobina, C. B. (2013). Single-phase power compensation in a current source converter. In *Proceedings of the IEEE ECCE*, Denver, CO (pp. 5288–5293).
6. Vitorino, M. A., Hartmann, L. V., Fernandes, D. A., et al. (2014). Single-phase current source converter with new modulation approach and power decoupling. In *Proceedings of the IEEE APEC*, Fort Worth, TX (pp. 2200–2207).
7. Liu, Y., Su, M., Liu, F., Zheng, M., Liang, X., Xu, G., & Sun, Y. (2019). Single-phase inverter with wide input voltage and power decoupling capability. *IEEE Access, 7*, 16870–16879.
8. Liu, Y., Sun, Y., & Su, M. (2017). A family of two-port switching networks with ripple power decoupling and output voltage step-up functions. *IET Power Electronics, 10*(10), 1175–1182.

# Virtual-Impedance-Based Decoupling Control

**6**

As mentioned in Chap. 1, an LC resonator tuned at twice the grid frequency can be inserted DC bus to block the ripple voltage caused by the ripple power [1]. However, this method requires large passive components as a result of relatively low resonate frequency, which will degrade the power density and even the reliability. What's worse, the decoupling performance will be deteriorated when the parameter drifts happen. In this chapter, a control strategy based on emulating the output voltage characteristic of the LC resonator is presented. Then, no bulky passive components are needed and the parameter drift issue is also avoided. In Sect. 6.1 the passive power decoupling using an LC resonant circuit applied for SCSR is analyzed detailly. Then, the active power control based on emulating LC resonator is introduced in Sect. 6.2. Section 6.3 gives some simulation and experimental results to verify the proposed control strategy. The extended application in single-phase voltage source rectifier (SVSR) is briefly introduced in Sect. 6.4. Finally, Sect. 6.5 discusses the pros and cons of the virtual-impedance-based decoupling control.

## 6.1 Passive Power Decoupling Using an LC Resonant Circuit

Figure 6.1 redraws the circuit structure of the SCSR with an LC resonant circuit [3]. According to (1.9) $i_{dc}$ contains an unexpected secondary pulsating component in steady-state.

The equivalent impedance of the parallel LC resonator is

$$Z_d = j\omega L_d // \frac{1}{j\omega C_d} = \frac{j\omega L_d}{1 - \omega^2 L_d C_d} \tag{6.1}$$

For the DC component in $u_r$ (see (1.4) for its expression), $Z_d$ is equal to zero and the resonator can be viewed as short-circuit. While, for the AC component, when $L_d$ and $C_d$ suffice

© The Author(s), under exclusive license to Springer Nature Switzerland AG 2023    105
Y. Liu, *Active Power Decoupling Technology in Single-Phase Current-Source Converters*,
Synthesis Lectures on Power Electronics, https://doi.org/10.1007/978-3-031-21270-3_6

**Fig. 6.1** Single-phase
current-type rectifier with an
LC resonator

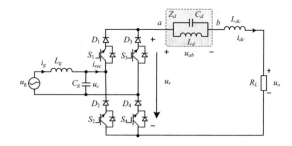

$$\omega_r = 1/\sqrt{L_d C_d} \tag{6.2}$$

$Z_d$ is equal to infinite and the resonator can be viewed as open-circuit. Then, the low frequency ripple voltage is prevented from entering the DC side. From the view of volt-second balance, the function of the LC resonator is to generate a voltage that is reversed with $\tilde{u}_r$. Then, the DC-link inductor is exempt from the secondary ripple voltage.

The relationship between the capacitance and the inductance when satisfying (6.2) is depicted in Fig. 6.2. Actually, if $C_d$ diminishes to zero, $L_d$ should be infinite, which is the case of employing an infinite large inductor $L_{dc}$. The method by employing the parallel LC resonator is easy to accomplish, but several defects also exist. Due to the low resonant frequency $2 f_g(100 \text{ Hz})$, passive components with large volume are unavoidably used. For instance, if $L_d$ takes 2.5 mH, $C_d$ needs 1000 μF, resulting in low power density. What's worse, the decoupling performance will deteriorate due to the tolerance or parameter drifts. To overcome these disadvantages, an active method of using electric circuits to emulate the output characteristic of the LC resonator is introduces in next section.

**Fig. 6.2** Relationship between
the capacitance and the
inductance using in the
physical LC resonator

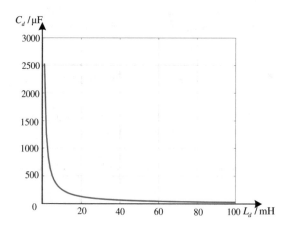

## 6.2 Active Power Control Based on Emulating LC Resonator

### 6.2.1 Basic Idea

According to previous analysis, the dynamic equation of $u_{ab}$ in Fig. 6.1 can be expressed as

$$u_{ab}(s) = \left( \frac{s/C_d}{s^2 + \omega_r^2} \right) i_{dc}(s) \tag{6.3}$$

In order to achieve the same external voltage characteristic of the LC resonator, an extra electric circuit is employed and its output voltage reference is designed as

$$u^*_{ab\_r}(s) = \left( \frac{k_r s}{s^2 + \omega_r^2} \right) i_{dc}(s) \tag{6.4}$$

where $k_r$ is the adjustable resonant gain.

In the ideal condition, the LC resonator only buffers the ripple power (reactive power) and will not consume power. However, to stabilize the capacitor voltage in the electric circuit, a resistance $R_d$ is virtualized to achieve voltage regulation. $R_d > 0$ ($R_d < 0$) denotes that the circuit absorbs power (releases power). Accordingly, the ultimate injected voltage reference is modified to

$$u^*_{ab}(s) = \underbrace{\left( R_d + \frac{k_r s}{s^2 + \omega_r^2} \right)}_{Z^*_d} i_{dc}(s) \tag{6.5}$$

### 6.2.2 Circuit Realization

Several feasible circuits to emulate the physical LC resonator are shown in Fig. 6.3. The terminal output voltage $u_{ab}$ is controlled to track the reference $u_{ab}^*$. Taking Fig. 6.3a as an example, when switches $S_5$ and $S_6$ are both turned off, the terminal output voltage $u_{ab}$ is equal to $u_d$; when switches $S_5$ and $S_6$ are both turned on, $u_{ab}$ is equal to $-u_d$; while only one switch is turned on, $u_{ab}$ becomes zero. These three voltage levels are used to synthesize the expected voltage $u_{ab}^*$. The emulator shown in Fig. 6.3a contains only one capacitor and two active switches, which can reduce the volume and the control complexity of the system. Therefore, it is adopted to verify the effectiveness of the proposed control method.

**Fig. 6.3** Feasible circuits to emulate the LC resonator. **a** Asymmetric H-bridge circuit. **b** H-bridge circuit

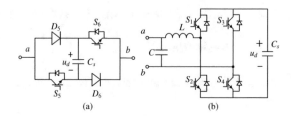

### 6.2.3  Control Strategy

The mathematical model can be established as follows

$$L_g \frac{di_g}{dt} = u_g - u_c \tag{6.6}$$

$$C_g \frac{du_c}{dt} = i_g - i_{rec} \tag{6.7}$$

$$L_{dc} \frac{di_{dc}}{dt} = u_r - u_{ab} - u_o \tag{6.8}$$

$$C_s \frac{du_d}{dt} = d_d i_{dc} \tag{6.9}$$

where $i_{rec} = d_r i_{dc}$, $u_r = d_r u_c$ and $u_{ab} = d_d u_d$. $d_r$ and $d_d$ are defined as follows:

$$\begin{cases} d_r = d_1 - d_2 \\ d_d = 1 - d_5 - d_6 \end{cases} \tag{6.10}$$

and $d_i$ ($i = 1, 2, 3, 4, 5, 6$) is the duty ratio of switch $S_i$.

The control diagram is depicted in Fig. 6.4. The rectifier is used to regulate the input current to achieve high power factor as well as transfer the power to the load. By multiplying $i_{dc}$ on both sides of (6.8), the following equation can be obtained,

$$\frac{L_{dc}}{2} \frac{di_{dc}^2}{dt} = u_c i_{rec} - \underbrace{d_r u_d i_{dc}}_{\tilde{p}} - \underbrace{u_o i_{dc}}_{P_o} \tag{6.11}$$

where $u_c i_{rec}$ implies the input power, $\tilde{p}$ is the secondary pulsating power and $P_o$ is the output power. Ignoring the effect of the LC low pass filter, $u_c$ is deemed to be equal to $u_g$. Suppose the input current reference is

$$i_{rec}^* = I^* \cos(\omega_g t) \tag{6.12}$$

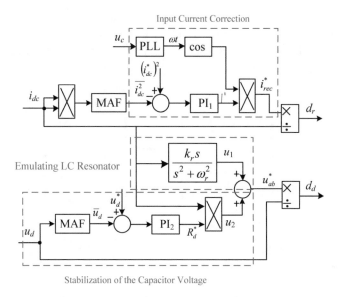

**Fig. 6.4** Block diagram of the control scheme

By substituting (6.12) into (6.11) and using the periodic average method [7], Eq. (6.11) turns into be

$$L_{dc}\frac{d\overline{i_{dc}^2}}{dt} = VI^* - 2P_o \tag{6.13}$$

The average value is obtained by using a moving average filter (MAF). To regulate the average value of $i_{dc}$, a PI controller is employed and $I^*$ is designed as

$$I^* = \frac{1}{V}\left\{\left[(i_{dc}^*)^2 - \overline{i_{dc}^2}\right]\left(k_{p1} + \frac{k_{i1}}{s}\right) + 2P_o\right\} \tag{6.14}$$

where $k_{p1}$ and $k_{i1}$ are the designed controller parameters.

The asymmetric H-bridge circuit is controlled to emulate the behavior of the LC resonator and stabilize the decoupling capacitor voltage. Based on (6.5), the reference capacitor voltage $u_{ab}^*$ is composed of two parts ($u_1$ and $u_2$ in the control block). $u_1$ is obtained by multiplying the DC current ($i_{dc}$) and the impedance of the LC resonator to achieve power decoupling. And $u_2$ is the product of the virtual resistance $R_d^*$ and the DC current ($i_{dc}$) to remain the average decoupling capacitor voltage. The virtual resistance $R_d^*$ is designed as

$$R_d^* = (u_d^* - \overline{u}_d)(k_{p2} + \frac{k_{i2}}{s}) \tag{6.15}$$

where $k_{p2}$ and $k_{i2}$ are the corresponding controller parameters, $u_d^*$ is the average capacitor voltage reference and $\bar{u}_d$ is achieved by using MAF. Once the control references $i_{rec}^*$ and $u_{ab}^*$ are obtained, the duty radios $d_i$ and $d_d$ can be easily calculated. Then, the duty ratio of each switch is designed as

$$
d_1 = \begin{cases} 1, & d_r > 0 \\ 0, & d_r \leq 0 \end{cases} \qquad d_2 = \begin{cases} 1 - d_r, & d_r > 0 \\ -d_r & d_r \leq 0 \end{cases}
$$

$$
d_3 = \begin{cases} 0, & d_r > 0 \\ 1, & d_r \leq 0 \end{cases} \qquad d_4 = \begin{cases} d_r, & d_r > 0 \\ 1 + d_r, & d_r \leq 0 \end{cases} \qquad (6.16)
$$

$$
d_5 = \begin{cases} 1 - d_d, & d_d > 0 \\ 1, & d_d \leq 0 \end{cases} \qquad d_6 = \begin{cases} 0, & d_d > 0 \\ -d_d, & d_d \leq 0 \end{cases}
$$

It is noted that for the LC emulator, the reference for achieving ripple power decoupling is derived only from $i_{dc}$. And this current sampling circuit can be embedded in the submodule of emulating LC resonator.

## 6.3    Simulation and Experiments Results

The proposed control method is verified by the numerical simulation and the physical test. The numerical simulation was carried out on the Matlab/Simulink platform. The main circuit includes a single-phase current source rectifier and an asymmetric H-bridge circuit, as shown in Fig. 6.3a. The asymmetric H-bridge circuit is in series with the DC-link inductor. In the experiment, semiconductors used are 1MBH60D-100 IGBTs and DSEI60-06A diodes. The control algorithm of the converter is realized by a combination of digital signal processor (DSP) TMS320F28335 and field programmable gate array (FPGA) EP2C8T144C8N. Some other specific parameters are listed in Table 6.1. The DC inductor current reference and the average capacitor voltage reference are set to 6 A and 120 V, respectively.

**Table 6.1** Parameters of the experimental setup (for SCSR)

| Symbol | Description | Value |
|--------|-------------|-------|
| $u_g$ | Input grid voltage | 110 V(rms) |
| $f_g$ | Input frequency | 50 Hz |
| $L_g$ | Input filter inductance | 0.6 mH |
| $C_g$ | Input filter capacitance | 20 μF |
| $C_s$ | Stored-energy capacitance | 100 μF |
| $L_{dc}$ | DC inductance | 3.6 mH |
| $R_L$ | Load resistance | 6 Ω |
| $f_s$ | Switching frequency | 20 kHz |

Figure 6.5 shows the steady-state waveforms of the DC current $i_{dc}$, the decoupling capacitor voltage $u_d$, the grid voltage $u_g$, and the grid current $i_g$. Experimental results in Fig. 6.5b match well with those in the simulation in Fig. 6.5a. The grid current $i_g$ is sine shaped and in phase with the grid voltage $u_g$. In addition, the load current $i_{dc}$ keeps constant at the given reference, which indicates that an excellent decoupling control is achieved. For the capacitor voltage $u_d$, it swings at twice the grid frequency as a result of buffering the ripple power. The measured efficiencies are 84 and 87% with/without the decoupling circuit. This issue can be mitigated by using wide-bandgap (WBG) semiconductor devices like the silicon carbide (SiC) and the gallium nitride (GaN)

To further manifest the decoupling effect of the proposed method, a test with disabling the proposed control method has been carried out. It is realized by setting the resonant frequency of the PR controller deviating from $200\pi$ rad/s (100 Hz) to $100\pi$ rad/s (50 Hz) on purpose. The dynamic waveforms of the simulation and experiments are respectively shown in Fig. 6.6a, b. The detuning of the resonator leads to very low impedance for the current $i_{dc}$. In this case, the output $u_1$ is nearly zero and the emulation of LC resonator with 100 Hz resonant frequency fails. Therefore, the load current $i_{dc}$ begins to oscillate and even falls down to zero at the valley. And the decoupling capacitor voltage is near flat and kept at 120 V due to that $u_2$ still works. The simulation and experimental results manifest the decoupling effect of the proposed method.

**Fig. 6.5** Steady-state waveforms of grid voltage $u_g$, grid current $i_g$, capacitor voltage $u_d$ and load current $i_{dc}$. **a** Simulation results. **b** Experimental results

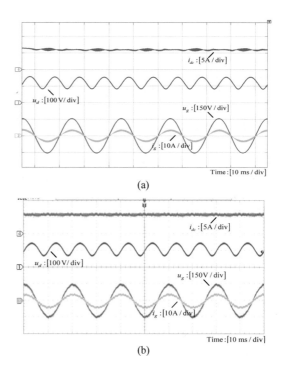

**Fig. 6.6** Dynamic waveforms when changing the resonant frequency. **a** Simulation results. **b** Experimental results

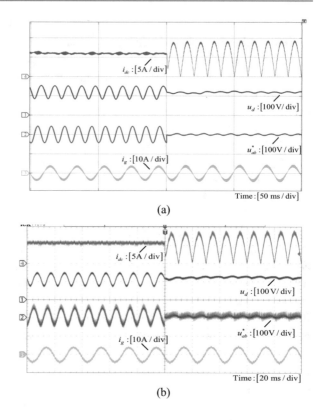

(a)

(b)

## 6.4    Extended Application in VSC

### 6.4.1    Working Principle

The virtual-impedance-based control method can also be applied to VSC [9, 10, 11]. Figure 6.7 redraws the main circuit of a single-phase voltage-source rectifier with a passive series LC resonator, which is directly connected in parallel to the DC bus as the independent decoupling circuit.

The equivalent conductance of the series LC resonator is

$$Y_d = \frac{1}{j\omega L_d + 1/(j\omega C_d)} = \frac{j\omega C_d}{1 - \omega^2 L_d C_d} \tag{6.17}$$

where $L_d$ and $C_d$ also suffice (6.2) and $\omega_r = 2\pi f_g$.

For the constant DC current in $i_r$ (see (1.5) for its expression), $Y_d = 0$, which means that the DC current is blocked. For the 2-order ripple current in $i_r$, $Y_d = \infty$, which means that $\tilde{i}_r$ is shorted. That is to say, $\tilde{i}_r$ totally flows through the resonant circuit.

The practical current $i_{ab}$ flowing through the series LC resonator can be expressed as

**Fig. 6.7** Single-phase voltage-source rectifier with passive series LC resonator

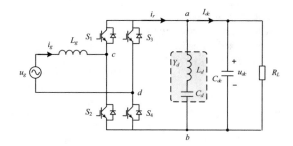

$$i_{ab}(s) = \left( \frac{s/L_d}{s^2 + \omega_r^2} \right) u_{dc}(s). \tag{6.18}$$

When adopting an electric circuit to construct the same impedance characteristics, the terminal current reference of the emulated LC resonator should be set to

$$i_{ab\_r}^*(s) = \left( \frac{k_r s}{s^2 + \omega_r^2} \right) u_{dc}(s). \tag{6.19}$$

In order to compensate for the power losses in the circuit, a virtual conductance $G_d$ is needed, as shown in Fig. 6.8. Likewise, $G_d > 0$ and $G_d < 0$ respectively means that the circuit absorbs active power and releases active power. And in the steady state, $G_d$ will be close to zero. Therefore, the terminal current reference is rewritten as

$$i_{ab}^*(s) = \underbrace{\left( G_d + \frac{k_r s}{s^2 + \omega_r^2} \right)}_{Y_d^*} u_{dc}(s) \tag{6.20}$$

The Buck circuit is taken as an example to achieve the control. The system circuit diagram is shown in Fig. 6.9, where $S_a$ and $S_b$ are the switches, and $C_s$ and $L_s$ are the decoupling capacitor and decoupling inductor.

According to Fig. 6.9, the mathematical model of the system can be established as follows.

**Fig. 6.8** Impedance characteristics construction of the decoupling circuit

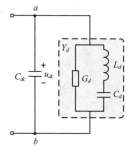

**Fig. 6.9** Single-phase
voltage-source rectifier with
the emulated series LC
resonator

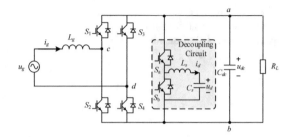

$$L_g \frac{di_g}{dt} = u_g - d_r u_{dc} \tag{6.21}$$

$$C_{dc} \frac{du_{dc}}{dt} = d_r i_g - d_d i_d - \frac{u_{dc}}{R_L} \tag{6.22}$$

$$L_s \frac{di_d}{dt} = d_d u_{dc} - u_d \tag{6.23}$$

$$C_s \frac{du_d}{dt} = i_d \tag{6.24}$$

where $d_r = d_1 - d_3$ is the duty cycle of the rectifier bridge ($d_1$ and $d_3$ are the duty cycles of switches $S_1$ and $S_3$), $d_d = d_{Sa}$ ($d_{Sa}$ is the duty cycle of switch $S_a$) is the duty cycle of the decoupling circuit.

For the whole system, the control contains two parts: the rectifier control and the decoupling control. The rectifier control is to regulate the average value of the DC bus voltage and the decoupling control is to realize power decoupling. The control block diagram is shown in Fig. 6.10.

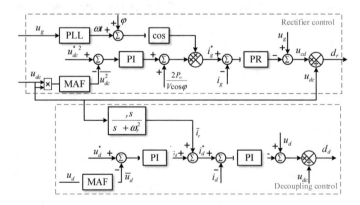

**Fig. 6.10** Control block diagram of virtual-impedance-based decoupling control applying in VSC

The adopted rectifier control is the classical double closed-loop control containing an outer voltage loop and an inner current loop. The outer voltage loop regulates the average DC bus voltage and its error is processed by the PI controller and then the output is used to modify the magnitude reference of AC current. The phase reference is obtained through a PLL. As for the inner current loop, a PR controller is used to track the sinusoidal current reference.

In terms of the decoupling control, a double closed-loop control structure is also used. According to (6.20), the current reference of the inner loop consists of two parts: the ripple current and the extra compensation current to maintain the decoupling capacitor voltage. The power losses of the buck decoupling circuit will pull down the capacitor voltage $u_d$. Therefore, the extra compensation current is obtained as

$$i_{Gd} = u_{dc}G_d = (u_d^* - \bar{u}_d)\left(k_p + \frac{k_i}{s}\right) \tag{6.25}$$

where $k_p$ and $k_i$ are the parameters of PI controller, $u_d^*$ is the average capacitor voltage reference and $\bar{u}_d$ is achieved by using MAF.

From Fig. 6.10, it can be seen that the decoupling control only needs to sample the DC bus voltage and the variables in the decoupling module. Meanwhile, from the perspective of the control structure, the decoupling control and rectifier control are completely independent and do not interfere with each other, thus there is no need for a unified central controller, which makes the realization of the plug-and-play function possible.

## 6.4.2 Experimental Results

To better verify the effect of the proposed control strategy, relevant experiments were conducted. The hardware platform consists of a single-phase voltage-source rectifier with 500 W output power and a buck-type decoupling circuit. The corresponding circuit parameters are listed in Table 6.2. And the references of the DC-bus voltage and the average decoupling capacitor voltage are 400 V and 300 V respectively.

Figure 6.11 shows the corresponding waveforms when the control strategy is not activated. They include the grid voltage $u_g$, the grid current $i_g$ and DC-bus voltage $u_{dc}$. As seen, $i_g$ is sinusoidal and stays in phase with $u_g$, and the average value of the DC bus voltage $u_{dc}$ is stable at 400 V. However, due to the small capacitor $C_{dc}$ (its value is 20 μF), the fluctuation of $u_{dc}$ ($\Delta v$) reaches about 200 V. The voltage ripple rate is up to 50% ($\Delta v/\bar{u}dc$), which is far from the actual industrial production requirements.

Figure 6.12 shows the experimental waveforms with enabling the control strategy. Figure 6.12a presents the steady state waveforms. It can be found that $i_g$ remains sinusoidal, but the fluctuation of $u_{dc}$ ($\Delta v$) is significantly reduced to be only 20 V. And the voltage ripple rate is 5%. The average value of $u_d$ is stabilized at 300 V. Figure 6.12b shows the transient experimental waveforms. When the decoupling circuit is connected at

**Table 6.2** Parameters of the experimental setup (for SVSR)

| Symbol | Description | Value |
|--------|-------------|-------|
| $u_g$ | Input grid voltage | 110 V(rms) |
| $f_g$ | Input frequency | 50 Hz |
| $L_g$ | Input filter inductance | 3 mH |
| $L_s$ | Decoupling inductance | 1.5 mH |
| $C_s$ | Decoupling capacitance | 100 μF |
| $C_{dc}$ | DC capacitance | 20 μF |
| $R_L$ | Load resistance | 320 Ω |
| $f_s$ | Switching frequency | 20 kHz |

**Fig. 6.11** Experimental waveforms of steady-state without decoupling circuit

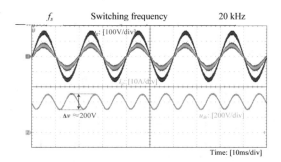

moment $t$, the voltage fluctuation of $u_{dc}$ starts to decrease and the decoupling capacitor voltage $u_d$ starts to fluctuate, which demonstrates that the 2-order ripple power transfers from the DC bus to the decoupling capacitor. However, for the ac side, $u_g$ and $i_g$ keep consistent before and after moment $t$, indicating that the decoupling circuit does not affect the rectifier operation. This performance can facilitate the realization of the plug-and-play.

## 6.5    Discussion

In this section, comparisons have been carried out among the proposed control method, the passive dimpling method, and existing active power decoupling control methods.

When a real LC resonator is used [18], the volume is usually large due to the low resonant frequency. For example, considering the parameters $L_d = 2.5$ mH and $C_d = 1000$ μF, the total volume is estimated to be 282.23 cm$^3$. The volume is estimated by adopting the ways in [2] and the capacitors from the manufacturer United Chemi-Con are adopted. However, the total volume is only 24.22 cm$^3$ with the proposed method [22]. On the other hand, due to the increased semiconductor devices, the reliability of the proposed method is reduced. The failure rate has been analyzed using the handbook

**Fig. 6.12** Experimental
waveforms with decoupling
circuit operation. **a** Steady state
results. **b** Transient state results

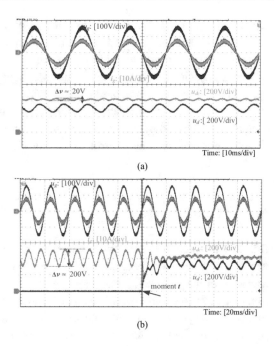

(a)

(b)

for reliability prediction of electronic equipment MIL- HDBK-217F [8, 12]. The results are 0.42 Failures/$10^6$ h for the passive dimpling method and 1.01 Failures/$10^6$ h for the proposed method.

The control methods are generally divided into two kinds, named Kinds A and B, according to whether the control facilitates the modularization of the decoupling circuit. In Kind A, the ripple power decoupling and the decoupling capacitor voltage stabilization are both achieved with only controlling the decoupling circuit (LC emulator in this paper) and only knowing the external information of the DC-link current/voltage. There is no information to be exchanged between the decoupling circuit and rectifier/inverter, which facilitates the modularization design of the decoupling circuit. The other control methods belong to Kind B. Most of the exiting control methods belong to Kind B, for examples, the ripple current compensation control (similar to the principle of active power filter) [15], the energy conservation based decoupling capacitor voltage tracking control [19], and the automatic-power-decoupling control [16, 21]. In this kind of control, a central controller is preferred to coordinate the operations of the decoupling circuit and the rectifier/inverter. It should be noted that to reduce component count, there are a lot of switch-multiplexing decoupling circuits [4, 20], in which there exist switches playing twofold roles of power decoupling and power conversion. For this kind of decoupling circuit, the designed control methods are all belong to Kind B. For the control of Kind A, the decoupling circuit is usually controlled to mimic a physical component, for examples, a capacitor [5, 14, 17], an inductor [6, 13, 17] and an LC resonator tuned at twice the grid frequency in this paper. In [5, 6, 13, 14], a pure differential operation is needed and the high-frequency noises will be amplified. This issue can be overcome by directly sampling the inductor voltage or capacitor current [17] and adding an inner controller. However, the inductor voltage or capacitor current is non-smooth which may increase the complexity of sampled signal processing. In the proposed method, both differential operation and voltage/current sensing network are avoided, and only the inductor current or capacitor voltage is utilized to emulate the desired dynamics. Therefore, the proposed method can be a good candidate for active power decoupling control, especially when the modular design of the decoupling circuit is expected.

## 6.6   Conclusion

This chapter proposes a control strategy to realize power decoupling in single-phase current source rectifier. By employing an electric circuit to obtain the same external characteristics of the passive LC resonator, the output power of double grid frequency is well buffered. From the view of volt-second balance, the proposed circuit generates an AC voltage to compensate the pulsating component and therefore the ripple voltage of the DC side is eliminated. The experimental results verify the effectiveness of emulating LC resonator well. In addition, the proposed idea can be extended to the single-phase voltage source converter.

## References

1. Singh, B., Singh, B. N., Chandra, A., Al-Haddad, K., Pandey, A., & Kothari, D. P. (2003). A review of single-phase improved power quality AC-DC converters. *IEEE Transactions on Industrial Electronics, 50*(5), 962–981.
2. Zhao, D., Liu, W., Shen, K., Zhao, G., & Wang, X. (2018). Multi-objective optimal de-sign of passive power filter for aircraft starter/generator system application. *The Journal of Engineering, 2018*(13), 636–641.
3. Han, H., Liu, Y., Sun, Y., Su, M., & Xiong, W. (2015). Single-phase current source converter with power decoupling capability using a series-connected active buffer. *IET Power Electronics, 8*(5), 700–707.
4. Hu, H., Harb, S., Kutkut, N., Batarseh, I., & Shen, Z. J. (2013). A review of power de-coupling techniques for microinverters with three different decoupling capacitor locations in PV systems. *IEEE Transactions on Power Electronics, 28*(6), 2711–2726.
5. Wang, H., & Wang, H. (2017). A two-terminal active capacitor. *IEEE Transactions on Power Electronics, 32*(8), 5893–5896.
6. Wang, H., & Wang, H. (2019). A two-terminal active inductor with minimum apparent power for the auxiliary circuit. *IEEE Transactions on Power Electronics, 34*(2), 1013–1016.
7. Sanders, J. A., Verhulst, F., & Murdock, J. (1985). *Averaging methods in nonlinear dynamical systems* (2nd ed.). Springer.

8. Soon, J. L., Lu, D. D. C., Peng, J. C., & Xiao, W. (2020). Reconfigurable non-isolated DC-DC converter with fault-tolerant capability. *IEEE Transactions on Power Electronics*, early access.
9. Fukushima, K., Norigoe, I., Shoyama, M., Ninomiya, T., Harada, Y., & Tsukakoshi, K. (2009). Input current-ripple consideration for the pulse-link DC-AC converter for fuel cells by small series LC circuit. In *2009 Twenty-Fourth Annual IEEE Applied Power Electronics Conference and Exposition*, Washington, DC (pp. 447–451).
10. Hwu, K., Tu, W., & Lai, C. (2014). Light-emitting diode driver with low-frequency ripple suppressed and dimming efficiency improved. *IET Power Electronics, 7*(1), 105–113.
11. Vasiladiotis, M., & Rufer, A. (2014). Dynamic analysis and state feedback voltage control of single-phase active rectifiers with DC-link resonant filters. *IEEE Transactions on Power Electronics, 29*(10), 5620–5633.
12. Military Handbook Reliability Prediction of Electronic Equipment, MILHDBK-217F, Dec. 2, 1991, Notice 2, Dept. Defense, Washington, D.C, USA, Feb. 28, 1995.
13. Brooks, N. C., Qin, S., & Pilawa-Podgurski, R. C. N. (2017). Design of an active power pulsation buffer using an equivalent series-resonant impedance model. In *2017 IEEE 18th Workshop on Control and Modeling for Power Electronics (COMPEL)*, Stanford, CA (pp. 1–7).
14. Chen, R., Liu, Y., & Peng, F. Z. (2017). A solid state variable capacitor with minimum capacitor. *IEEE Transactions on Power Electronics, 32*(7), 5035–5044.
15. Wang, R., et al. (2011). A high power density single-phase PWM rectifier with active ripple energy storage. *IEEE Transactions on Power Electronics, 26*(5), 1430–1443.
16. Li, S., Qi, W., Tan, S.-C., & Hui, S. Y. (2018). Enhanced automatic-power-decoupling control method for single-phase AC-to-DC converters. *IEEE Trans-actions on Power Electronics, 33*(2), 1816–1828.
17. Li, S., Qi, W., Tan, S.-C., Hui, S. Y., & Tse, C. K. (2018). A general approach to programmable and reconfigurable emulation of power impedances. *IEEE Transactions on Power Electronics, 33*(1), 259–271.
18. Nonaka, S., & Neba, Y. (1993). Single-phase PWM current source converter with double-frequency parallel resonance circuit for DC smoothing. In *Conference Record, Industry Applications Society, IEEE-IAS Annual Meeting* (pp. 1144–1151).
19. Ohnuma, Y., & Itoh, J. (2014). A novel single-phase Buck PFC AC–DC converter with power decoupling capability using an active buffer. *IEEE Transactions on Industry Applications, 50*(3), 1905–1914.
20. Sun, Y., Liu, Y., Su, M., Xiong, W., & Yang, J. (2016). Review of active power decoupling topologies in single-phase systems. *IEEE Transactions on Power Electronics, 31*(7), 4778–4794.
21. Sun, Y., Liu, Y., Su, M., Li, X., & Yang, J. (2016). Active power decoupling method for single-phase current source rectifier with no additional active switches. *IEEE Transactions on Power Electronics, 31*(8), 5644–5654.
22. Liu, Y., Zhang, W., Lin, J., Su, M., & Liang, X. (2021). Active power decoupling control for single-phase current source rectifier based on emulating LC resonator. *IEEE Transactions on Industrial Electronics, 68*(6), 5460–5465.

# Stability Analysis and Improvement Based on HSS Modeling

**7**

Single-phase AC-DC converters with decoupling circuit is a non-linear time-periodic (NLTP) system. The classical tools, such as root locus and Routh stability criterion, which are widely used in linear system, cannot be directly applied. In this chapter, the harmonic state-space (HSS) modeling method is adopted to analyze the stability of the SCSR with a power decoupling circuit under constant power loads (CPLs). By making the best use of the decoupling circuit, a simple and general method of adding a modulation term to the control output reference is proposed to improve the system stability. Section 7.1 gives a brief revisit of the HSS. Section 7.2 demonstrates the application of HSS in SCSR with a decoupling circuit. Analytical and simulation results are provided in Sect. 7.3.

## 7.1    Brief Revisit of HSS Modeling

Recently, HSS modeling method, which is based on the linear time-periodic (LTP) theory, receives much attention for analyzing the harmonic coupling and stability of power-electronic-based systems. In the harmonic domain, this method considers the continuous dynamics of the power systems and the switching dynamics of power electronic converters, which contributes to the improvement of analysis accuracy. Moreover, the HSS model is capable of discovering the harmonic transfer mechanism in both the ac-side and dc-side and the harmonic interaction. Furthermore, this modeling method provides an accurate and simple way to analyze the stability of single-phase systems due to the fact that many stability analysis tools developed for common LTI systems can be generalized to LTP systems, such as eigenvalue loci analysis and Nyquist stability criterion.

Consider a general continuous NLTP system which is $T$-periodic

Y. Liu, *Active Power Decoupling Technology in Single-Phase Current-Source Converters*, Synthesis Lectures on Power Electronics, https://doi.org/10.1007/978-3-031-21270-3_7

$$\begin{cases} \dot{x}(t) = f(x(t), u(t)) \\ y(t) = g(x(t), u(t)) \end{cases} \quad (7.1)$$

where $x \in R^n$ is the state vector of the system, $u \in R^q$ is the input, and $y \in R^p$ is the output. The steady-state trajectories of state and input signal satisfy the time-periodicity $x_0(t + T) = x_0(t)$ and $u_0(t + T) = u_0(t)$. Therefore, this NLTP system can be linearized around its steady-state trajectory to obtain an LTP system

$$\begin{cases} \dot{\tilde{x}}(t) = A(t)\tilde{x}(t) + B(t)\tilde{u}(t) \\ \tilde{y}(t) = C(t)\tilde{x}(t) + D(t)\tilde{u}(t) \end{cases} \quad (7.2)$$

where

$$A(t) = \frac{\partial f}{\partial x}(x_0(t), u_0(t)), \quad B(t) = \frac{\partial f}{\partial u}(x_0(t), u_0(t))$$

$$C(t) = \frac{\partial g}{\partial x}(x_0(t), u_0(t)), \quad D(t) = \frac{\partial g}{\partial u}(x_0(t), u_0(t)) \quad (7.3)$$

are all $T$-periodic matrices.

To develop the frequency response of the LTP system, the exponentially modulated periodic (EMP) test signal is used, which is defined as

$$\tilde{u}(t) = \sum_{n=-\infty}^{\infty} \tilde{U}_k e^{st} e^{jk\omega_T t} \quad (s = j\omega) \quad (7.4)$$

where $\tilde{U}_k$ is the $k$th Fourier coefficient and $\omega_T = 2\pi/T$. By injecting this signal into the LTP system, and applying the harmonic balance theory, this system in Fourier matrix expansion form can be derived as

$$\begin{cases} (s + jk\omega_T)\tilde{X}_k = \sum_{m=-\infty}^{\infty} A_{k-m}\tilde{X}_m + \sum_{m=-\infty}^{\infty} B_{k-m}\tilde{U}_m \\ \tilde{Y}_k = \sum_{m=-\infty}^{\infty} C_{k-m}\tilde{X}_m + \sum_{m=-\infty}^{\infty} D_{k-m}\tilde{U}_m \end{cases} \quad (7.5)$$

These equations precisely describe the input-state-output relationship in the Fourier coefficients domain. However, the convolution of the complex Fourier series is complicated, so the Toeplitz transformation is introduced to simplify this procedure. The Toeplitz transformation maps the set of complex Fourier coefficients into a doubly infinite block Toeplitz matrix, which is defined as

$$T[A(t)] = \mathcal{A} = \begin{bmatrix} \ddots & \vdots & \vdots & \vdots & \cdot{\cdot}^{\cdot} \\ \cdots & A_0 & A_{-1} & A_{-2} & \cdots \\ \cdots & A_1 & A_0 & A_{-1} & \cdots \\ \cdots & A_2 & A_1 & A_0 & \cdots \\ \cdot{\cdot}^{\cdot} & \vdots & \vdots & \vdots & \ddots \end{bmatrix} \tag{7.6}$$

where $A_i$ represents the Fourier matrix coefficient of $A(t)$.

Hence, the system Eq. (7.5) can be rewritten in the Toeplitz form as

$$\begin{cases} s\mathcal{X} = (\mathcal{A} - \mathcal{N})\mathcal{X} + \mathcal{B}\mathcal{U} \\ \mathcal{Y} = \mathcal{C}\mathcal{X} + \mathcal{D}\mathcal{U} \end{cases} \tag{7.7}$$

which is called the HSS model of the LTP system, where $\mathcal{N} = \text{blkdiag}\{jk\omega_T E\}$ is a block diagonal matrix and $E$ denotes the identity matrix with the same dimensions as $A(t)$.

Based on the HSS model, the stability of this system can be determined by the eigenvalues of the matrix $(\mathcal{A} - \mathcal{N})$. If all the eigenvalues lie on the left of the complex plane or those lying on the imaginary axis have algebraic multiplicity equal to 1, the system is stable, otherwise, the system is unstable.

Clearly, in the practical implementation of stability analysis, these Toeplitz matrices must be truncated in which the truncation order refers to the maximum number of harmonics taken into account. It is found that the truncation error would be negligible if the truncation order is greater than a certain value.

## 7.2 Application of HSS in SCSR with the Decoupling Circuit

### 7.2.1 HSS Modeling

The investigated SCSR with the decoupling circuit is depicted in Fig. 7.1. It is introduced in detail in Chap. 3. The only difference is that a CPL rather than a resistor is connected.

A closed-loop power decoupling control scheme is developed as shown in Fig. 7.2. In this scheme, the phase angle of input voltage $u_c$ is obtained through a second-order generalized integrator (SOGI)-phase-locked loop (PLL). For the control of CSR, a notch filter is adopted to obtain the average voltage $\bar{u}_d$ of the decoupling capacitor, which is used to generate the reference amplitude $I_{rec}^*$ of input current through a PI controller. Also, the input current of CSR is synchronized with input voltage and displacement angle $\varphi$ is usually set to zero to achieve the unity power factor. In the decoupling control loop, the load current regulator output and feedforward terms are used for calculating the reference voltage $u_{ab}^*$ of ABC. And $\gamma_{stable}$ is the stabilization term that will be introduced in the next section.

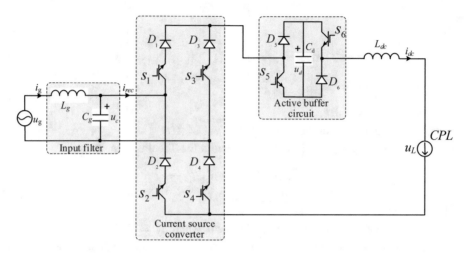

**Fig. 7.1** Single-phase current source rectifier feeding an ideal CPL

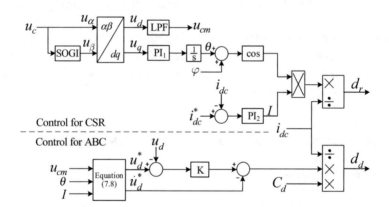

**Fig. 7.2** Block diagram of the control scheme

In the decoupling control loop, the output of the PLL and load current regulator output are used for calculating the reference of active buffer capacitor voltage $u_d^*$ and its corresponding derivative $\dot{u}_d^*$ by the following equations

$$
\begin{cases}
u_d^* = \sqrt{\bar{u}_d^2 + \dfrac{u_{cm} I \sin(2\omega_g t + \varphi)}{2\omega_g C_d}} \\
\dot{u}_d^* = \dfrac{u_{cm} I \cos(2\omega_g t + \varphi)}{2C_d u_d^*}
\end{cases}
\tag{7.8}
$$

where $\omega_g$ is the angular frequency of source voltage and is a given constant, which should satisfy the following constraint

$$\bar{u}_d \geq \sqrt{\frac{u_{cm}I}{2\omega_g C_d}} \qquad (7.9)$$

In this control scheme, one low-pass filter (LPF), one SOGI that introduces a $T_g/4$ delay at $\omega_g$, and three PI controllers are adopted, whose transfer functions are given as

$$LPF(s) = \frac{1}{1 + \tau_f s} \qquad (7.10)$$

$$SOGI(s) = \frac{2\xi\omega_g^2}{s^2 + 2\xi\omega_g s + \omega_g^2} \qquad (7.11)$$

$$PI_j(s) = k_{pj} + \frac{k_{ij}}{s} \ (j = 1, \ 2) \qquad (7.12)$$

where $\tau_f$ is the time constant, $\xi$ is the damping factor, and $k_{pj}$ and $k_{ij}$ are corresponding controller parameters.

The whole system model is a ten-order continuous NLTP model, which is represented in the state-space form

$$\begin{cases} \dot{x}_1 = x_2 \\ \dot{x}_2 = -\omega_g^2 x_1 - 2\xi\omega_g x_2 + 2\xi\omega_g^2 x_8 \\ \dot{x}_3 = x_4 - k_{p1}\sin(x_3)x_8 + k_{p1}\cos(x_3)x_1 \\ \dot{x}_4 = -k_{i1}\sin(x_3)x_8 + k_{i1}\cos(x_3)x_1 \\ \dot{x}_5 = [x_8\cos(x_3) + x_1\sin(x_3) - x_5]/\tau_f \\ \dot{x}_6 = i_{dc}^* - x_9 \\ \dot{x}_7 = (u_g - x_8)/L_g \\ \dot{x}_8 = (x_7 - d_r x_9)/C_g \\ \dot{x}_9 = \left(d_r x_8 - d_d x_{10} - \frac{P^*}{x_9}\right)/L_{dc} \\ \dot{x}_{10} = (d_d x_9)/C_d \end{cases} \qquad (7.13)$$

where

$$\begin{cases} I = k_{p2}(i_{dc}^* - x_9) + k_{i2}x_6 \\ d_r = [I\cos(x_3)]/x_9 \\ u_d^* = \sqrt{\bar{u}_d^2 + \frac{Ix_5\sin(2x_3)}{2\omega_g C_d}} \\ \dot{u}_d^* = [Ix_5\cos(2x_3)]/(2C_d u_d^*) \\ d_d = [C_d(\dot{u}_d^* + K(u_d^* - x_{10}))]/x_9 \end{cases} \qquad (7.14)$$

In this model, $x_1$–$x_5$ describe the internal dynamics of the SOGI-PLL, where $x_1$ is the virtual in-quadrature component $u_\beta$ of the input voltage, $x_3$ is the estimated phase angle $\theta$ and $x_5$ is the filtered voltage amplitude $u_{cm}$; $x_6$ is the internal dynamic of output PI regulator; $x_7$–$x_{10}$ represent the grid current $i_g$, the capacitor voltage $u_c$, the load current $i_{dc}$, and the active buffer capacitor voltage $u_d$, respectively. The excitation source of this system is the grid voltage $u_g$. And $P_{CPL}$ represents the required power of CPL.

To linearize the NLTP system, the steady-state trajectories should be solved. Through some mathematical manipulation, the sets of steady-state solutions are given as

$$\begin{cases} \bar{x}_1 = \bar{x}_5 \sin(\omega_g t + \angle \bar{x}_8) \\ \bar{x}_2 = \omega_g \bar{x}_5 \cos(\omega_g t + \angle \bar{x}_8) \\ \bar{x}_3 = \omega_g t + \angle \bar{x}_8 \\ \bar{x}_4 = \omega_g \\ \bar{x}_5 = |\bar{x}_8| \end{cases} \tag{7.15-I}$$

$$\begin{cases} \bar{x}_6 = 2P^* / (k_{i2}\bar{x}_5) \\ \bar{x}_7 = \left[2P^* \cos(\bar{x}_3)\right]/\bar{x}_5 - C_g \omega_g \bar{x}_5 \sin(\bar{x}_3) \\ \bar{x}_8 = \bar{x}_5 \cos(\bar{x}_3) \\ \bar{x}_9 = i_{dc}^* \\ \bar{x}_{10} = \sqrt{\bar{u}_d^2 + \dfrac{P^* \sin(2\bar{x}_3)}{\omega_g C_d}} \end{cases} \tag{7.15-II}$$

The steady-state solutions are presented in an analytical form, but in some special instances, the NLTP systems can only be solved numerically.

Following the steps illustrated in the previous section, linearization around the steady-state solution is applied to the system model (7.13), then the small-signal LTP model can be derived in the form of $\dot{\tilde{x}}(t) = A(t)\tilde{x}(t)$ where $A(t)$ is given as

$$A(t) = \begin{bmatrix} 0 & 1 & 0 & 0 & 0 & 0 & 0 & 0 & 0 & 0 \\ -\omega_g^2 & -2\xi\omega_g & 0 & 0 & 0 & 0 & 0 & 2\xi\omega_g^2 & 0 & 0 \\ k_{p1}\cos(\bar{x}_3) & 0 & -k_{p1}\left[\cos(\bar{x}_3)\bar{x}_8 + \sin(\bar{x}_3)\bar{x}_1\right] & 1 & 0 & 0 & 0 & -k_{p1}\sin(\bar{x}_3) & 0 & 0 \\ k_{p1}\cos(\bar{x}_3) & 0 & -k_{i1}\left[\cos(\bar{x}_3)\bar{x}_8 + \sin(\bar{x}_3)\bar{x}_1\right] & 0 & 0 & 0 & 0 & -k_{i1}\sin(\bar{x}_3) & 0 & 0 \\ \dfrac{\sin(\bar{x}_3)}{\tau_f} & 0 & \dfrac{\cos(\bar{x}_3)\bar{x}_1 - \sin(\bar{x}_3)\bar{x}_8}{\tau_f} & 0 & -\dfrac{1}{\tau_f} & 0 & 0 & \dfrac{\cos(\bar{x}_3)}{\tau_f} & 0 & 0 \\ 0 & 0 & 0 & 0 & 0 & 0 & 0 & 0 & -1 & 0 \\ 0 & 0 & 0 & 0 & 0 & 0 & 0 & \dfrac{-1}{L_g} & 0 & 0 \\ 0 & 0 & \dfrac{2P^* \sin(\bar{x}_3)}{C_g \bar{x}_5} & 0 & 0 & -\dfrac{k_{i2}\cos(\bar{x}_3)}{C_g} & \dfrac{1}{C_g} & 0 & \dfrac{k_{p2}\cos(\bar{x}_3)}{C_g} & 0 \\ 0 & 0 & \dfrac{\bar{\mu}_1}{L_{dc}} & 0 & \dfrac{\bar{\mu}_2}{L_{dc}} & \dfrac{\bar{\mu}_3}{L_{dc}} & 0 & \dfrac{2P^* \cos(\bar{x}_3)}{i_{dc}^* \bar{x}_5 L_{dc}} & \dfrac{\bar{\mu}_4}{L_{dc}} & \dfrac{\bar{\mu}_5}{L_{dc}} \\ 0 & 0 & \bar{\eta}_1 & 0 & \bar{\eta}_2 & \bar{\eta}_3 & 0 & 0 & \bar{\eta}_4 & -K \end{bmatrix} \tag{7.16}$$

where

$$
\begin{cases}
\overline{\mu}_0 = \left[ C_d \left(\overline{u}_d^*\right)^2 K - P^* \cos(2\overline{x}_3) \right] \Big/ \left( 4\omega_g C_d \left(\overline{u}_d^*\right)^2 i_{dc}^* \right) \\[4pt]
\overline{\mu}_1 = -4P^* \left( \overline{\mu}_0 \cos(2\overline{x}_3) - \frac{\sin(2\overline{x}_3)}{2 i_{dc}^*} \right) - \left[ P^* \sin(2\overline{x}_3) \right] \Big/ i_{dc}^* \\[4pt]
\overline{\mu}_2 = -\frac{2P^*}{\overline{x}_5} \left( \overline{\mu}_0 \sin(2\overline{x}_3) + \frac{\cos(2\overline{x}_3)}{2 i_{dc}^*} \right) \\[4pt]
\overline{\mu}_3 = -k_{i2}\overline{x}_5 \left( \overline{\mu}_0 \sin(2\overline{x}_3) + \frac{\cos(2\overline{x}_3)}{2 i_{dc}^*} \right) + \left[ k_{i2}\overline{x}_5 [\cos(2\overline{x}_3) + 1] \right] \Big/ \left( 2 i_{dc}^* \right) \\[4pt]
\overline{\mu}_4 = k_{p2}\overline{x}_5 \left( \overline{\mu}_0 \sin(2\overline{x}_3) + \frac{\cos(2\overline{x}_3)}{2 i_{dc}^*} \right) - \left[ k_{p2}\overline{x}_5 [\cos(2\overline{x}_3) + 1] \right] \Big/ \left( 2 i_{dc}^* \right) \\[4pt]
\overline{\mu}_5 = \left[ K C_d \overline{u}_d^* \right] \Big/ i_{dc}^* - \left[ P^* \cos(2\overline{x}_3) \right] \Big/ \left( \overline{u}_d^* i_{dc}^* \right)
\end{cases}
\tag{7.17-I}
$$

$$
\begin{cases}
\overline{\eta}_0 = \left[ C_d \left(\overline{u}_d^*\right)^2 K - P^* \cos(2\overline{x}_3) \right] \Big/ \left[ 4\omega_g C_d^2 \left(\overline{u}_d^*\right)^3 \right] \\[4pt]
\overline{\eta}_1 = 4P^* \left( \overline{\eta}_0 \cos(2\overline{x}_3) - \frac{\sin(2\overline{x}_3)}{2 C_d \overline{u}_d^*} \right) \\[4pt]
\overline{\eta}_2 = \frac{2P^*}{\overline{x}_5} \left( \overline{\eta}_0 \sin(2\overline{x}_3) + \frac{\cos(2\overline{x}_3)}{2 C_d \overline{u}_d^*} \right) \\[4pt]
\overline{\eta}_3 = k_{i2}\overline{x}_5 \left( \overline{\eta}_0 \sin(2\overline{x}_3) + \frac{\cos(2\overline{x}_3)}{2 C_d \overline{u}_d^*} \right) \\[4pt]
\overline{\eta}_4 = -k_{p2}\overline{x}_5 \left( \overline{\eta}_0 \sin(2\overline{x}_3) + \frac{\cos(2\overline{x}_3)}{2 C_d \overline{u}_d^*} \right)
\end{cases}
\tag{7.17-II}
$$

The Toeplitz transform is applied to $A(t)$ and the system stability can be determined by the eigenvalues of $(\mathcal{A} - \mathcal{N})$. Unlike the system whose state variables are only DC or single-frequency $T_g$-periodic AC, there are more than one operating trajectory in this system and the state variables are strong coupled in the $A(t)$, so that the corresponding matrix coefficients of the Fourier series expansion only can be solved in a numerical way.

## 7.2.2  Stabilization Method

Most literature about stability analysis of single-phase systems is focused on the factors leading to instability, but fewer methods are given for stability improvement. It is found that the ABC provides a freedom for improving system stability. Therefore, the control strategy of decoupling control loop is modified as

$$
\begin{cases}
u_d^* = \sqrt{\overline{u}_d^2 + \frac{u_{cm} I \sin(2x_3)}{2\omega C_d}} + \gamma_{stable} \\[6pt]
\dot{u}_d^* = \dfrac{u_{cm} I \cos(2x_3)}{2 C_d \sqrt{\overline{u}_d^2 + \frac{u_{cm} I \sin(2x_3)}{2\omega C_d}}}
\end{cases}
\tag{7.18}
$$

where $\gamma_{stable}$ is a modification term that should satisfy some constraints.

Inspired by the idea of the constructive method, a simple and general method to construct the modification terms is proposed. First, the basic requirements of the construction method are stated as follows:

(1) The actual output is equal to its reference when entering the steady-state, i.e. that $\gamma_{stable}$ should be constant at a steady state so that the resulting deviation could be eliminated by the PI controller;
(2) The resulting deviation during the transient state must be small enough;
(3) The stability boundary must be broadened. Due to the negative impedance characteristic of CPL, this condition can be described as

$$Re\left\{\left.\frac{\partial \gamma_{stable}}{\partial i_{dc}}\right|_{i_{dc}=i_{dc}^*}\right\} > 0 \qquad (7.19)$$

which is equivalent to visualizing a positive impedance to attenuate the adverse impacts of CPL.

According to the previous requirements, several intuitively constructive examples are given as follows.

Case 1:

$$\gamma_{stable1} = k_f i_{dc} \qquad (7.20)$$

Case 2:

$$\gamma_{stable2} = k_f \mathcal{L}^{-1}\left\{\frac{\tau_s s}{1+\tau_s s} i_{dc}(s)\right\} \qquad (7.21)$$

Case 3:

$$\gamma_{stable3} = k_f \left(e^{\alpha \cdot i_{dc}} - 1\right) \qquad (7.22)$$

where $k_f$ is the stabilization gain ($k_f > 0$), $L^{-1}\{\}$ represents the inverse Laplace transformation, $\tau_s$ is the time constant of the high-pass filter and $\alpha$ is the amplification coefficient ($\alpha > 0$). As seen, all these examples comply with all the principles mentioned above when adjusted parameters are selected properly. Taking Case 1 as an example, a proper $k_f$ can guarantee stability which is verified by analytical and simulation results in the following section. However, if $k_f$ is too large, the modification term would cause large pulsating power transferring to the dc side and degrade the system response speed. Thus, the selection of $k_f$ is a tradeoff between stability and dynamic performance.

To present an analytical result of case 1, the modified linearized model is derived where only the following two terms should be replaced in (7.16)

$$\begin{cases} \overline{\mu}_4^m = \overline{\mu}_4 - \left[C_d K k_f \overline{u}_d^*\right]/i_{dc}^* \\ \overline{\eta}_4^m = \overline{\eta}_4 + K k_f \end{cases} \qquad (7.23)$$

The other two cases can be analyzed in the same way, but the derivation of the corresponding linearized model is not reported for the sake of brevity.

## 7.3    Analytical and Simulation Results

### 7.3.1    Continuous LTP Analytical Results

To verify the validity of the proposed scheme, simulations are performed in MAT-LAB/Simulink. The specifications of this system are summarized in Table 7.1. The stability analysis in this subsection will focus on the required power $P^*$ of CPL.

To implement the eigenvalue analysis, a proper truncation order should be chosen. Therefore, considering a high precision, the Toeplitz matrix of $A(t)$ is truncated at $N = 100$.

In the first, the eigenvalue analysis is performed on the linearized model (7.16) without using the stabilization method. As Fig. 7.3 shows, all the eigenvalues lie on the left-half plane with $P^* = 225$ W, which denotes the system is stable. But this system is unstable when $P^* = 230$ W since some eigenvalues are shifted to the right-half plane. This conforms to the fact that the system tends to be unstable as the required power of CPL increases.

To verify the proposed stabilization method, the linearized model (7.23) related to Case 1 is used to plot the corresponding eigenvalue loci. In this analysis, $k_f = 2$ and $P^* = 230$ W are applied. Under this condition, no unstable eigenvalue exists in Fig. 7.4, indicating that this method has a beneficial impact on system stability.

**Table 7.1**  The system parameters

| Symbol | Description | Value |
|---|---|---|
| $u_g$ | Input phase voltage | 90 V |
| $\omega_g$ | Input angular frequency | 314 rad/s |
| $L_g$ | Input filter inductance | 0.6 mH |
| $C_g$ | Input filter capacitance | 20 μF |
| $L_{dc}$ | Output inductance | 3 mH |
| $C_d$ | Active buffer capacitance | 91.8 μF |
| $\bar{u}_d$ | Decoupling control param | 100 V |
| $K$ | Decoupling control param | 30 |
| $i_{dc}^*$ | DC current reference | 6 A |
| $k_{p1}$ | PLL PI param | 27.207 |
| $k_{i1}$ | PLL PI param | 493.48 |
| $\xi$ | PLL SOGI param | 0.707 |
| $\tau_f$ | PLL LPF param | 0.0159 |
| $k_{p2}$ | Output PI param | 0.084 |
| $k_{i2}$ | Output PI param | 98.7 |

**Fig. 7.3** Eigenvalue loci without stabilization method. **a** $P^* = 225$ W, **b** Zoom of (**a**) around the imaginary axis, **c** $P^* = 230$ W, **d** Zoom of (**c**) around the imaginary axis

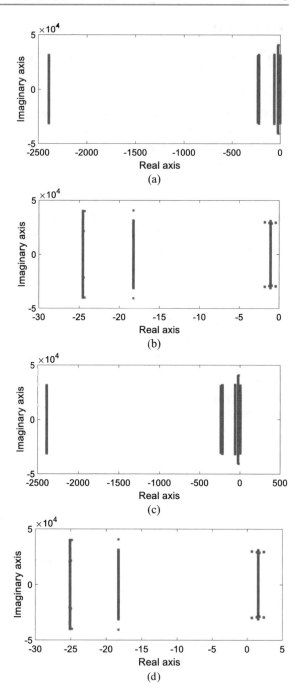

**Fig. 7.4** Eigenvalue loci with stabilization method (Case 1: $k_f = 2$). **a** $P^* = 230$ W. **b** Zoom of (**a**) around the imaginary axis

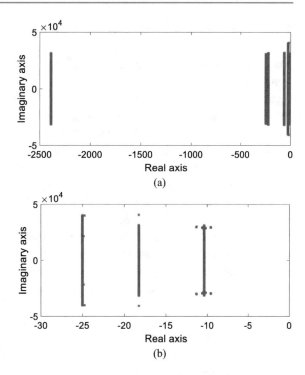

Also, the influence of $k_f$ is discussed by detecting the stability boundary with different $k_f$. The area below the curve represents the stable operating area in Fig. 7.5. As observed, the stability region becomes larger with increasing $k_f$. And the stability boundary is at $P^* = 246.4$ W when $k_f = 2$.

**Fig. 7.5** Stability boundary of load power $P^*$ against different stabilization gain $k_f$

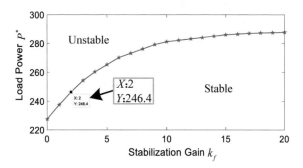

### 7.3.2    Time-Domain Simulation Results

The analysis is performed on the average model (7.13) to verify the assessed stability boundary. Figure 7.6 shows the waveforms of load current $i_{dc}$, load power $P^*$, grid current $i_g$, and active buffer capacitor voltage $u_d$. From 0.6 to 1.6 s, this system is stable under $P^* = 225$ W because the oscillated amplitude of load current is degraded gradually as Fig. 7.7a shows. After $t = 1.6$ s, the $P^*$ is changed to 230 W, then the load current tends to be divergent which is depicted in Fig. 7.7b. Besides, the grid current and active buffer capacitor voltage begin to oscillate. Finally, this system becomes unstable which is consistent with the above analysis results in Fig. 7.3.

Adopting the stabilization method of Case 1 with $k_f = 2$, the corresponding waveforms are presented in Fig. 7.8 and a detailed description of load current is depicted in Fig. 7.9. It is obvious that load current is convergent under $P^* = 245$ W before 1.5 s. However, after $t = 1.6$ s, the load current becomes oscillating when $P^*$ increases to 250 W. The stability boundary is enlarged from $P^* = 225$ W to $P^* = 245$ W. So, it can be concluded that the stabilization method is effective to broaden the system stability region.

## 7.4    Conclusion

This chapter establishes the HSS model of the SCSR with power decoupling capacity under CPL. Based on the developed model, the stability region related to the required power of CPL is assessed precisely. By making use of the power decoupling circuit, a stabilization method is proposed to improve system stability by constructing a modification term in the decoupling control loop. Also, the relationship between the stabilization gain and stability region is presented. Simulations are carried out to verify the effectiveness of the proposed method.

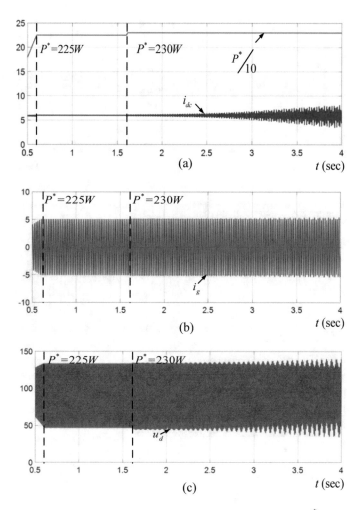

**Fig. 7.6** Simulation waveforms without stabilization method. **a** Load power $P^*$ divided by 10, Load current $i_{dc}$, **b** Grid current $i_g$, **c** Active buffer capacitor voltage $u_d$

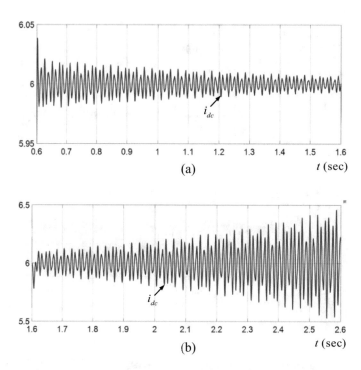

**Fig. 7.7**  Zoom of load current $i_{dc}$ waveform in Fig. 7.6. **a** Zoom of stable load current from 0.6 to 1.6 s, **b** Zoom of unstable load current from 1.6 to 2.6 s

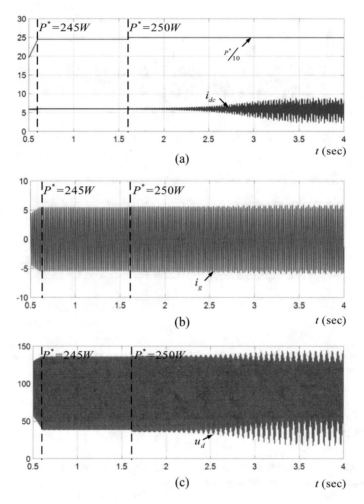

**Fig. 7.8** Simulation waveforms with stabilization method (Case 1: $k_f = 2$). **a** Load power $P^*$ divided by 10, Load current $i_{dc}$. **b** Grid current $i_g$. **c** Active buffer capacitor voltage $u_d$

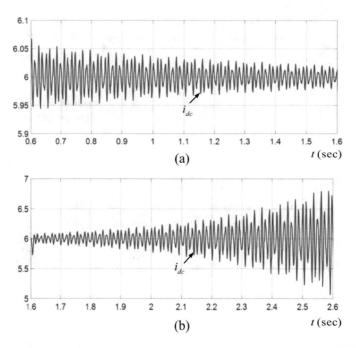

**Fig. 7.9** Zoom of load current $i_{dc}$ waveform in Fig. 7.8. **a** Zoom of stable load current from 0.6 to 1.6 s. **b** Zoom of unstable load current from 1.6 to 2.6 s

## Bibliography

1. Bohm, M. L., & Allgower, F. (2012). Stability of periodically timer-varying systems: Periodic Lyapunov functions. *Automatica, 48*(10), 2663–2669.
2. Emadi, A., Khaligh, C. H., & Rivetta and G. A. Williamson,. (2006). Constant power loads and negative impedance instability in automotive systems: Definition, modeling, stability, and control of power electronic converters and motor drives. *IEEE Transactions on Vehicular Technology, 55*(4), 1112–1125.
3. Han, H., Liu, Y., Sun, Y., Su, M., & Xiong, W. (2015). Single-phase current source converter with power de-coupling capability using a series-connected active buffer. *IET Power Electronics, 8*(5), 700–707.
4. Sandberg, H., Mollerstedt, E., & Bernhardsson,. (2005). Frequency-domain analysis of linear time-periodic systems. *IEEE Transactions on Automatic Control, 50*(12), 1971–1983.
5. Wang, H., Wu, M., & Sun, J. (2015). Analysis of low-frequency oscillation in electric railways based on small-signal modeling of vehicle-grid systems in DQ frame. *IEEE Transactions on Power Electronics, 30*(9), 5318–5330.
6. Kwon, J., Wang, X., Blaabjerg, F., Bak, C., Wood, A., & Watson, N. R. (2016). Harmonic instability analysis of a single-phase grid-connected converter using a harmonic state-space modeling method. *IEEE Transactions on Industry Applications, 52*(5), 4188–4200.

7. Kwon, J., Wang, X., Blaabjerg, F., Bak, C., Sularea, V.-S., & Busca, C. (2017). Harmonic interaction analysis in a grid-connected converter using harmonic state-space (HSS) modeling. *IEEE Transactions on Power Electronics, 32*(9), 6823–6835.

8. Rodriguez, J. R., Dixon, J. W., Espinoza, J. R., Pontt, J., & Lezana, P. (2005). PWM regenerative rectifiers: State of the art. *IEEE Transactions on Industrial Electronics, 52*(1), 5–22.

9. Sun, J., & Karimi, K. (2008). Small-signal input impedance modeling of line-frequency rectifiers. *IEEE Transactions on Aerospace and Electronic Systems, 44*(4), 1489–1497.

10. Kisacikoglu, M., Ozpineci, B., & Tolbert, L. (2013). EV/PHEV bidirectional charger assessment for V2G reactive power operation. *IEEE Transactions on Power Electronics, 28*(12), 5717–5727.

11. Mollerstedt. (2000). *Dynamic analysis of harmonics in electrical systems*. Ph.D. Dissertation, Department Automatic Control, Lund Institute of Technology, Lund, Sweden.

12. Wereley, N. M. (1991). *Analysis and control of linear periodically time varying systems*. Ph.D. Dissertation, Department of Aeronautics and Astronautics, MIT Institute of Technology.

13. Krein, P. T., Balog, R. S., & Mirjafari, M. (2012). Minimum energy and capacitance requirements for single-phase inverters and rectifiers using a ripple port. *IEEE Transactions on Power Electronics, 27*(11), 4690–4698.

14. Bittanti, S., & Colaneri, P. (2009). *Periodic systems: Filtering and control*. Springer-Verlag.

15. Lissandron, S., Santa, L., Mattavelli, P., & Wen, B. (2016). Experimental validation for impedance-based small-signal stability analysis of single-phase interconnected power systems with grid-feeding inverters. *IEEE Journal of Emerging and Selected Topics in Power Electronics, 4*(1), 103–115.

16. Singh, B. N., Singh, A., Chandra, K., Al-Haddad, A. P., & Kothari, D. P. (2003). A review of single-phase improved power quality AC-DC converters. *IEEE Transactions on Industrial Electronics, 50*(5), 962–981.

17. Salis, V., Costabeber, A., Cox, S. M., Zanchetta, P., & Formentini, A. (2017). Stability boundary analysis in single-phase grid-connected inverters with PLL by LTP theory. *IEEE Transactions on Power Electronics*, early access.

18. Salis, V., Costabeber, A., Cox, S. M., & Zanchetta, P. (2017). Stability assessment of power-converter-based AC systems by LTP theory: Eigenvalue analysis and harmonic impedance estimation. *IEEE Journal of Emerging and Selected Topics in Power Electronics, 5*(4), 1513–1525.

19. Wen, D., Boroyevich, R., Burgos, P. M., & Shen, Z. (2015). Small-signal stability analysis of three-phase ac systems in the presence of constant power loads based on measured d-q frame impedances. *IEEE Transactions on Power Electronics, 30*(10), 5952–5963.

20. Han, Y., Fang, X., Yang, P., Wang, C., Xu, L., & Guerrero, J. (2018). Stability analysis of digital controlled single-phase inverter with synchronous reference frame voltage control. *IEEE Transactions on Power Electronics, 33*(7), 6333–6350.

21. Sun, Y., Su, M., Li, X., Wang, H., & Gui, W. (2013). A general constructive approach to matrix converter stabilization. *IEEE Transactions on Power Electronics, 28*(1), 418–430.

22. Sun, Y., Liu, Y., Su, M., Xiong, W., & Yang, J. (2016). Review of active power decoupling topologies in single-phase systems. *IEEE Transactions on Power Electronics, 31*(7), 4778–4794.

23. Sun, Y., Liu, Y., Su, M., Li, X., & Yang, J. (2016). Active power decoupling method for single-phase current-source rectifier with no additional active switches. *IEEE Transactions on Power Electronics, 31*(8), 5644–5654.

24. Liu, Z., Su, M., Sun, Y., Han, H., Hou, X., & Guerrero, J. M. (2017). Stability analysis of DC microgrids with constant power load under distributed control methods. *Automatica*, early access.

# Stability Analysis Based on DHSS Modeling

**8**

Digitally controlled single-phase active power decoupling converter is a continuous discrete hybrid system with NLTP nature, thus the conventional LTI model cannot accurately describe the system dynamic behavior. The DHSS modeling method is suitable to analyze the stability of digitally controlled converter. First, the continuous discrete hybrid system is unified into a time-periodic discrete model. After linearization, the system matrix in the discrete model is extended to an infinite dimensional Toeplitz matrix to realize the transformation of the LTP system into the LTI system. Finally, the stability can be analyzed by the Jacobian matrix eigenvalues of DHSS model. Section 8.1 introduces the modeling process of the DHSS. Section 8.2 shows stability boundary by employing the eigenvalues loci. In Sect. 8.3, the experimental results verify the accuracy of the model.

## 8.1 DHSS Modeling of SCSR with a Decoupling Circuit

### 8.1.1 System Description

The schematic diagram of studied converter is shown in Fig. 8.1, which has been introduced in Chap. 3. It is mainly composed of SCSR, a series-connected active power decoupling circuit (APDC) and a resistor load. The APDC is composed of asymmetric H-bridge including two decoupling capacitors ($C_a$, $C_b$), one inductor ($L_1$) and two switches ($S_a$, $D_b$).

The digital control diagram is depicted in Fig. 8.2, comprised of an SOGI-PLL and two control loops. The SOGI-PLL is used for obtaining the grid voltage phase $\theta$. The current control loop regulates control variable $d_d$ to maintain the DC current $i_{dc}$ constant by a proportional controller $P_3(z)$ and a disturbance feed forward compensation. And the voltage loop regulates the variable $d_r$ to stabilize the average value of the sum of two

© The Author(s), under exclusive license to Springer Nature Switzerland AG 2023
Y. Liu, *Active Power Decoupling Technology in Single-Phase Current-Source Converters*,
Synthesis Lectures on Power Electronics, https://doi.org/10.1007/978-3-031-21270-3_8

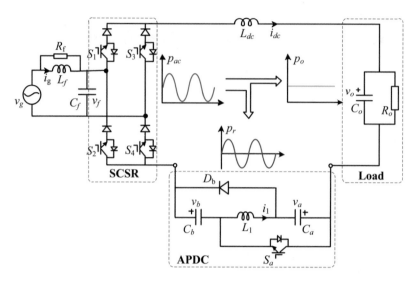

**Fig. 8.1**  Topology of the studied converter

decoupling capacitor voltages ($v_{com} = v_a + v_b$). Because the control algorithm is executed by the digital signal processing (DSP) chip, one step control delay here is considered, i.e., the duty cycles in $(k + 1)$th switching cycle are calculated in the $k$th switching cycle. It is worth noting that the relationships between the control variables ($d_r$, $d_d$) and the duty cycles satisfy

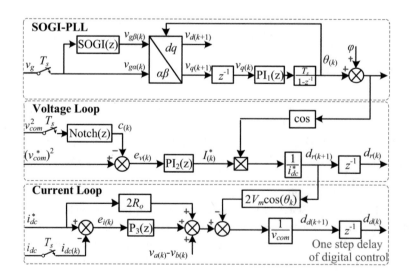

**Fig. 8.2**  Diagram of digital control for the studied converter

$$d_{c(k)} = d_{r(k)}\operatorname{sgn}(d_{r(k)}) \tag{8.1}$$

$$d_{a(k)} = \frac{1 + d_{d(k)}}{2} \tag{8.2}$$

where $d_c$ indicates the duty cycle of the high frequency switch $S_1$ or $S_3$ in the SCSR, and $d_a$ indicates the duty cycle of the switch $S_a$, and the subscript $(k)$ represents the $k$th switching cycle.

The waveforms of the DC inductor current during each operation mode are demonstrated in Fig. 8.3. As seen, each operation mode includes three subintervals, each subinterval represents a switching state, the system state space equation is given by

$$\dot{x}_c(t) = A_i x_c(t) + B v_g(t) \tag{8.3}$$

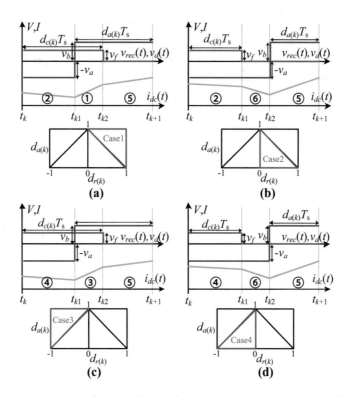

**Fig. 8.3** Key operational waveforms. **a** Case 1: $d_{a(k)} + d_{c(k)} > 1$ and $d_{r(k)} > 0$. **b** Case 2: $d_{a(k)} + d_{c(k)} < 1$ and $d_{r(k)} > 0$. **c** Case 3: $d_{a(k)} + d_{c(k)} > 1$ and $d_{r(k)} < 0$. **d** Case 4: $d_{a(k)} + d_{c(k)} < 1$ and $d_{r(k)} < 0$

where $\boldsymbol{x_c}(t) = [i_g, i_{dc}, i_1, v_f, v_a, v_b, v_o]^T$, $i = \{1, 2, \ldots, 6\}$, and the system matrix $\boldsymbol{A_i}$ at each switching state is listed in Table 8.1.

Moreover, the input matrix $\boldsymbol{B}$ is given by

$$\boldsymbol{B} = \begin{bmatrix} \frac{1}{L_f} & 0 & 0 & \frac{1}{R_f C_f} & 0 & 0 & 0 \end{bmatrix}^T \tag{8.4}$$

**Table 8.1** Switching states and system matrix $A_i$

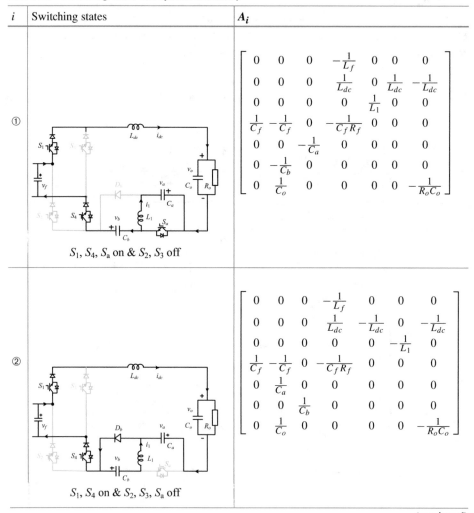

| $i$ | Switching states | $A_i$ |
|---|---|---|
| ① | $S_1$, $S_4$, $S_a$ on & $S_2$, $S_3$ off | $\begin{bmatrix} 0 & 0 & 0 & -\frac{1}{L_f} & 0 & 0 & 0 \\ 0 & 0 & 0 & \frac{1}{L_{dc}} & 0 & \frac{1}{L_{dc}} & -\frac{1}{L_{dc}} \\ 0 & 0 & 0 & 0 & \frac{1}{L_1} & 0 & 0 \\ \frac{1}{C_f} & -\frac{1}{C_f} & 0 & -\frac{1}{C_f R_f} & 0 & 0 & 0 \\ 0 & 0 & -\frac{1}{C_a} & 0 & 0 & 0 & 0 \\ 0 & -\frac{1}{C_b} & 0 & 0 & 0 & 0 & 0 \\ 0 & \frac{1}{C_o} & 0 & 0 & 0 & 0 & -\frac{1}{R_o C_o} \end{bmatrix}$ |
| ② | $S_1$, $S_4$ on & $S_2$, $S_3$, $S_a$ off | $\begin{bmatrix} 0 & 0 & 0 & -\frac{1}{L_f} & 0 & 0 & 0 \\ 0 & 0 & 0 & \frac{1}{L_{dc}} & -\frac{1}{L_{dc}} & 0 & -\frac{1}{L_{dc}} \\ 0 & 0 & 0 & 0 & 0 & -\frac{1}{L_1} & 0 \\ \frac{1}{C_f} & -\frac{1}{C_f} & 0 & -\frac{1}{C_f R_f} & 0 & 0 & 0 \\ 0 & \frac{1}{C_a} & 0 & 0 & 0 & 0 & 0 \\ 0 & 0 & \frac{1}{C_b} & 0 & 0 & 0 & 0 \\ 0 & \frac{1}{C_o} & 0 & 0 & 0 & 0 & -\frac{1}{R_o C_o} \end{bmatrix}$ |

(continued)

**Table 8.1** (continued)

| $i$ | Switching states | $A_i$ |
|---|---|---|

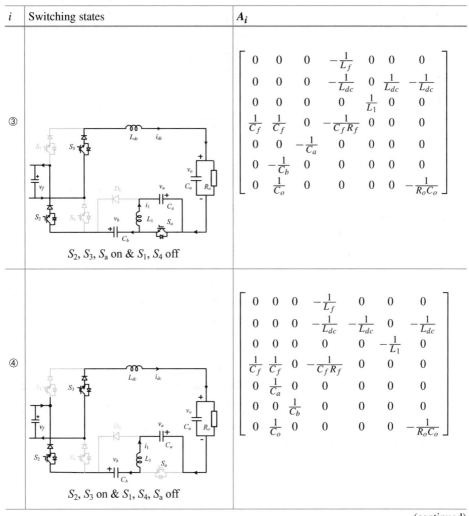

③ 

$S_2, S_3, S_a$ on & $S_1, S_4$ off

$$A_3 = \begin{bmatrix} 0 & 0 & 0 & -\frac{1}{L_f} & 0 & 0 & 0 \\ 0 & 0 & 0 & -\frac{1}{L_{dc}} & 0 & \frac{1}{L_{dc}} & -\frac{1}{L_{dc}} \\ 0 & 0 & 0 & 0 & \frac{1}{L_1} & 0 & 0 \\ \frac{1}{C_f} & \frac{1}{C_f} & 0 & -\frac{1}{C_f R_f} & 0 & 0 & 0 \\ 0 & 0 & -\frac{1}{C_a} & 0 & 0 & 0 & 0 \\ 0 & -\frac{1}{C_b} & 0 & 0 & 0 & 0 & 0 \\ 0 & \frac{1}{C_o} & 0 & 0 & 0 & 0 & -\frac{1}{R_o C_o} \end{bmatrix}$$

④ 

$S_2, S_3$ on & $S_1, S_4, S_a$ off

$$A_4 = \begin{bmatrix} 0 & 0 & 0 & -\frac{1}{L_f} & 0 & 0 & 0 \\ 0 & 0 & 0 & -\frac{1}{L_{dc}} & -\frac{1}{L_{dc}} & 0 & -\frac{1}{L_{dc}} \\ 0 & 0 & 0 & 0 & 0 & -\frac{1}{L_1} & 0 \\ \frac{1}{C_f} & \frac{1}{C_f} & 0 & -\frac{1}{C_f R_f} & 0 & 0 & 0 \\ 0 & \frac{1}{C_a} & 0 & 0 & 0 & 0 & 0 \\ 0 & 0 & \frac{1}{C_b} & 0 & 0 & 0 & 0 \\ 0 & \frac{1}{C_o} & 0 & 0 & 0 & 0 & -\frac{1}{R_o C_o} \end{bmatrix}$$

(continued)

## 8.1.2 Discrete State Space (DSS) Model of Main Circuit

Assumed that the switching transients are negligible and the circuit during each switching state is linear, the state transfer equation can be derived by

$$x_c(t) = f_i(x_c(t_0), t_s) = \Phi_i(t, t_0)x_c(t_0) + \psi_i(t, t_0) \tag{8.5}$$

where $f_i$ represents the iterative function during $i$th switching state, $t_s$ represents duration of $i$th switching state, and the state transition matrix (STM) $\Phi_i$ and $\psi_i$ are given by

**Table 8.1** (continued)

| $i$ | Switching states | $A_i$ |
|---|---|---|
| ⑤ | 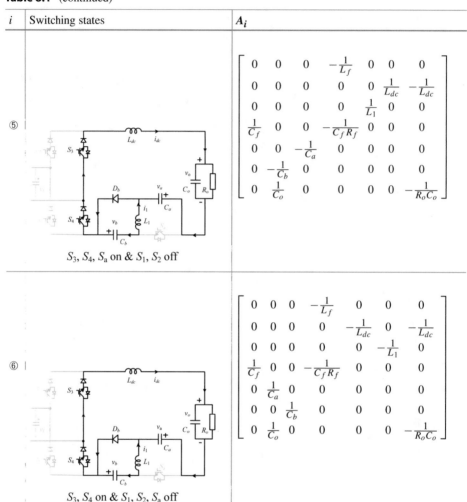  $S_3, S_4, S_a$ on & $S_1, S_2$ off | $\begin{bmatrix} 0 & 0 & 0 & -\frac{1}{L_f} & 0 & 0 & 0 \\ 0 & 0 & 0 & 0 & 0 & \frac{1}{L_{dc}} & -\frac{1}{L_{dc}} \\ 0 & 0 & 0 & 0 & \frac{1}{L_1} & 0 & 0 \\ \frac{1}{C_f} & 0 & 0 & -\frac{1}{C_f R_f} & 0 & 0 & 0 \\ 0 & 0 & -\frac{1}{C_a} & 0 & 0 & 0 & 0 \\ 0 & -\frac{1}{C_b} & 0 & 0 & 0 & 0 & 0 \\ 0 & \frac{1}{C_o} & 0 & 0 & 0 & 0 & -\frac{1}{R_o C_o} \end{bmatrix}$ |
| ⑥ | $S_3, S_4$ on & $S_1, S_2, S_a$ off | $\begin{bmatrix} 0 & 0 & 0 & -\frac{1}{L_f} & 0 & 0 & 0 \\ 0 & 0 & 0 & 0 & -\frac{1}{L_{dc}} & 0 & -\frac{1}{L_{dc}} \\ 0 & 0 & 0 & 0 & 0 & -\frac{1}{L_1} & 0 \\ \frac{1}{C_f} & 0 & 0 & -\frac{1}{C_f R_f} & 0 & 0 & 0 \\ 0 & \frac{1}{C_a} & 0 & 0 & 0 & 0 & 0 \\ 0 & 0 & \frac{1}{C_b} & 0 & 0 & 0 & 0 \\ 0 & \frac{1}{C_o} & 0 & 0 & 0 & 0 & -\frac{1}{R_o C_o} \end{bmatrix}$ |

$$\boldsymbol{\Phi}_i(t, t_0) = e^{A_i t_s} = e^{A_i(t-t_0)} \tag{8.6}$$

$$\boldsymbol{\psi}_i(t, t_0) = \int_{t_0}^{t} \boldsymbol{\Phi}_i(t, \tau)\boldsymbol{B}v_g(\tau)d\tau \tag{8.7}$$

Since the $v_g(\tau)$ is continuous periodic excitation, the system is nonautonomous. To accurately calculate the integral item $\boldsymbol{\psi}_i$, $v_g(\tau)$ is expanded by Euler's formula, i.e.,

$$v_g(\tau) = V_m \cos(\omega\tau) = V_m \frac{e^{j\omega\tau} + e^{-j\omega\tau}}{2} \tag{8.8}$$

Then substituting (8.6) and (8.8) into the integral item (8.7) leads to

$$\boldsymbol{\psi}_i(t, t_0) = \frac{V_m}{A_i^2 + \omega^2 \boldsymbol{I}} [\omega \boldsymbol{I} \sin(\omega t) - A_i \cos(\omega t)$$

$$-\omega e^{A_i(t-t_0)} \sin(\omega t_0) + A_i e^{A_i(t-t_0)} \cos(\omega t_0)] \boldsymbol{B} \tag{8.9}$$

For easy understanding, taking the case 1 $(d_{a(k)} + d_{c(k)} > 1$ and $d_{r(k)} > 0)$ as an example, the iteration process is shown in Fig. 8.4. And the iteration relation during $k$th switching cycle can be expressed as

$$\begin{aligned}
\boldsymbol{x}_{c(k1)} &= \boldsymbol{f}_2(\boldsymbol{x}_{c(k)}, t_{s1}) = \boldsymbol{\Phi}_2(t_{k1}, t_k)\boldsymbol{x}_{c(k)} + \boldsymbol{\psi}_2(t_{k1}, t_k) \\
\boldsymbol{x}_{c(k2)} &= \boldsymbol{f}_1(\boldsymbol{x}_{c(k1)}, t_{s2}) = \boldsymbol{\Phi}_1(t_{k2}, t_{k1})\boldsymbol{x}_{c(k1)} + \boldsymbol{\psi}_1(t_{k2}, t_{k1}) \\
\boldsymbol{x}_{c(k+1)} &= \boldsymbol{f}_5(\boldsymbol{x}_{c(k2)}, t_{s3}) = \boldsymbol{\Phi}_5(t_{k+1}, t_{k2})\boldsymbol{x}_{c(k2)} + \boldsymbol{\psi}_5(t_{k+1}, t_{k2})
\end{aligned} \tag{8.10}$$

where $t_k$ is the beginning time of $k$th switching cycle, $\boldsymbol{x}_{c(k1)}$ and $\boldsymbol{x}_{c(k2)}$ represent the values of $\boldsymbol{x}_c$ at the switching times $t_{k1}$ and $t_{k2}$ respectively, $\boldsymbol{x}_{c(k)}$ and $\boldsymbol{x}_{c(k+1)}$ represent the initial values of $\boldsymbol{x}_c$ at $k$th and $(k + 1)$th switching cycles respectively, and durations of three subintervals are given by

$$\begin{aligned}
t_{s1} &= t_{k1} - t_k = (1 - d_{a(k)})T_s \\
t_{s2} &= t_{k2} - t_{k1} = (d_{a(k)} + d_{r(k)} - 1)T_s \\
t_{s3} &= t_{k+1} - t_{k2} = (1 - d_{r(k)})T_s
\end{aligned} \tag{8.11}$$

According to (8.10) and Fig. 8.4, the DSS model from $\boldsymbol{x}_{c(k)}$ to $\boldsymbol{x}_{c(k+1)}$ can be expressed as

$$\begin{aligned}
\boldsymbol{x}_{c(k+1)} &= \boldsymbol{f}_5(\boldsymbol{f}_1(\boldsymbol{f}_2(\boldsymbol{x}_{c(k)}, t_{s1}), t_{s2}), t_{s3}) \\
&= \boldsymbol{\Phi}_5(t_{k+1}, t_{k2})\{\boldsymbol{\Phi}_1(t_{k2}, t_{k1})[\boldsymbol{\Phi}_2(t_{k1}, t_k)\boldsymbol{x}_{c(k)}
\end{aligned}$$

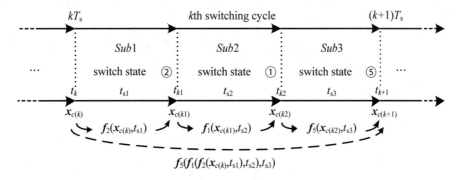

**Fig. 8.4** Iteration process of state vector under case 1

$$+\boldsymbol{\psi}_2(t_{k1}, t_k)\big] + \boldsymbol{\psi}_1(t_{k2}, t_{k1})\big\} + \boldsymbol{\psi}_5(t_{k+1}, t_{k2})$$
$$= \boldsymbol{G}_{c(k)}\boldsymbol{x}_{c(k)} + \boldsymbol{H}_{c(k)} \tag{8.12}$$

where

$$\boldsymbol{G}_{c(k)} = \boldsymbol{\Phi}_5(t_{k+1}, t_{k2})\boldsymbol{\Phi}_1(t_{k1}, t_k)\boldsymbol{\Phi}_2(t_{k2}, t_{k1})$$
$$\boldsymbol{H}_{c(k)} = \boldsymbol{\Phi}_5(t_{k+1}, t_{k2})\boldsymbol{\Phi}_1(t_{k2}, t_{k1})\boldsymbol{\psi}_2(t_{k1}, t_k)$$
$$+ \boldsymbol{\Phi}_5(t_{k+1}, t_{k2})\boldsymbol{\psi}_1(t_{k2}, t_{k1}) + \boldsymbol{\psi}_5(t_{k+1}, t_{k2}) \tag{8.13}$$

The iteration processes of the remaining three cases are similar to (8.12).

### 8.1.3   DSS Model of Digital Controller

The digital SOGI-PLL is used for tracking the phase $\theta_{(k)}$ of the grid voltage $v_g$. It generates a signal $v_{g\beta(k)}$ lagging the grid voltage by $\pi/2$, and its continuous transfer function is represented as

$$\mathrm{SOGI}(s) = \frac{2\zeta\omega_l^2}{s^2 + 2\zeta\omega_l s + \omega_l^2} \tag{8.14}$$

where $\omega_l$ is the angular frequency of grid voltage, and $\zeta$ is the damping ratio of the SOGI, which is directly proportional to the dynamic response of the system and inversely proportional to the filtering effect.

Considering the switching period $T_s$ as the discrete sample period, the bilinear transformation between s-domain and z-domain is given by

$$s = \frac{2}{T_s}\frac{z-1}{z+1} \tag{8.15}$$

Substituting (8.15) into (8.14), the discrete transfer function of SOGI-PLL is represented as

$$\mathrm{SOGI}(z) = \frac{b_0(1 + 2z^{-1} + z^{-2})}{1 + a_1 z^{-1} + a_0 z^{-2}} \tag{8.16}$$

where

$$a_0 = \frac{\omega_l^2 T_s^2 - 4\xi\omega_l T_s + 4}{\omega_l^2 T_s^2 + 4\xi\omega_l T_s + 4}$$

$$a_1 = \frac{2\omega_l^2 T_s^2 - 8}{\omega_l^2 T_s^2 + 4\xi\omega_l T_s + 4} \tag{8.17}$$

$$b_0 = \frac{2\xi\omega_l^2 T_s^2}{\omega_l^2 T_s^2 + 4\xi\omega_l T_s + 4}$$

According to the linear transformation of controllable canonical form, the DSS model of (8.16) is given as

$$\begin{bmatrix} x_8 \\ x_9 \end{bmatrix}_{(k+1)} = \begin{bmatrix} 0 & -a_0 \\ 1 & -a_1 \end{bmatrix}\begin{bmatrix} x_8 \\ x_9 \end{bmatrix}_{(k)} + \begin{bmatrix} b_0(1-a_0) \\ b_0(2-a_1) \end{bmatrix}v_{g(k)}$$

$$v_{g\beta(k)} = \begin{bmatrix} 0 & 1 \end{bmatrix}\begin{bmatrix} x_8 \\ x_9 \end{bmatrix}_{(k)} + b_0 v_{g(k)}$$

$$v_{g\alpha(k)} = v_{g(k)} \tag{8.18}$$

Then, applying the Park transformation to $v_{g\alpha}$ and $v_{g\beta}$, the q-axis voltage $v_{q(k+1)}$ is obtained

$$v_{q(k+1)} = \cos\big[\theta_{(k)}\big]v_{g\beta(k)} - \sin\big[\theta_{(k)}\big]v_{g\alpha(k)} \tag{8.19}$$

Perturbing the variables in the time-domain, the linearized form of (8.19) is derived as

$$\hat{v}_{q(k+1)} = \cos\big[\theta_{(k)}\big]\hat{v}_{g\beta(k)} - \sin\big[\theta_{(k)}\big]\hat{v}_{g\alpha(k)} - V_m\hat{\theta}_{(k)} \tag{8.20}$$

Based on the SOGI-PLL diagram, the following relationship satisfies

$$\theta_{(k)} = \mathcal{Z}\left[\frac{PI_1(s)}{s}\right]v_{q(k)} = \frac{k_{p1}T_s + k_{i1}T_s^2 - k_{p1}T_s z^{-1}}{1 - 2z^{-1} + z^{-2}}v_{q(k)} \tag{8.21}$$

Similarly, the DSS model of (8.21) is given by

$$\begin{bmatrix} x_{10} \\ x_{11} \end{bmatrix}_{(k+1)} = \begin{bmatrix} 0 & -1 \\ 1 & 2 \end{bmatrix}\begin{bmatrix} x_{10} \\ x_{11} \end{bmatrix}_{(k)} + \begin{bmatrix} -k_{p1}T_s - k_{i1}T_s^2 \\ k_{p1}T_s + 2k_{i1}T_s^2 \end{bmatrix}v_{q(k)}$$

$$\theta_{(k)} = \begin{bmatrix} 0 & 1 \end{bmatrix}\begin{bmatrix} x_{10} \\ x_{11} \end{bmatrix}_{(k)} + \big(k_{p1}T_s + k_{i1}T_s^2\big)v_{q(k)} \tag{8.22}$$

Typically, the digital notch filter is used to mitigate the second-order harmonic in the voltage loop, its discrete transfer function is given by

$$\text{Notch}(z) = \frac{c(z)}{r(z)} = \frac{(1+\alpha)}{2} \frac{1 - 2\beta z^{-1} + z^{-2}}{1 - \beta(1+\alpha)z^{-1} + \alpha z^{-2}} \tag{8.23}$$

and $\alpha$ and $\beta$ are represented as

$$\alpha = \frac{1 - \tan(\pi f_c/f_s)}{1 + \tan(\pi f_c/f_s)}$$

$$\beta = \cos(4\pi f_l/f_s) \tag{8.24}$$

where $f_l$ is the power frequency, $f_c$ is the central bandwidth, and $f_s$ is the discrete sample frequency.

Similarly, the DSS model of (8.23) is given by

$$\begin{bmatrix} x_{13} \\ x_{14} \end{bmatrix}_{(k+1)} = \begin{bmatrix} 0 & -\alpha \\ 1 & \beta(1+\alpha) \end{bmatrix} \begin{bmatrix} x_{13} \\ x_{14} \end{bmatrix}_{(k)} + \begin{bmatrix} \frac{1-\alpha^2}{2} \\ \frac{\alpha^2-1}{2}\beta \end{bmatrix} r_{(k)}$$

$$c_{(k)} = \begin{bmatrix} 0 & 1 \end{bmatrix} \begin{bmatrix} x_{13} \\ x_{14} \end{bmatrix}_{(k)} + \frac{1+\alpha}{2} r_{(k)} \tag{8.25}$$

where $r_{(k)}$ and $c_{(k)}$ are the input and the output of the notch filter in $k$th step respectively, and the $r_{(k)}$ is represented as

$$r_{(k)} = v_{com(k)}^2 = \left(v_{a(k)} + v_{b(k)}\right)^2 \tag{8.26}$$

As shown in the control diagram of the Fig. 8.2, the controller samples the current $i_{dc(k)}$ and the voltage $c_{(k)}$ at the beginning time $t_k = kT_s$ of $k$th switching cycle and then the error signals $(e_{i(k)}, e_{v(k)})$ is obtained. Furthermore, the error signals go through the digital controller to generate the control variables $(d_{r(k+1)}, d_{d(k+1)})$, and then the DSP calculates and updates the duty cycles $(d_{c(k+1)}, d_{a(k+1)})$ at the beginning time $t_{k+1} = (k + 1)T_s$ of the next switching cycle. Based on this, the DSS model of the digital controller is represented as

$$d_{d(k+1)} = \frac{1}{v_{com(k)}} \left\{ k_{p3}e_{i(k)} + v_{a(k)} - v_{b(k)} + 2i_{dc}^* R_o - 2d_{r(k+1)} V_m \cos[\theta_{(k)}] \right\}$$

$$d_{r(k+1)} = \frac{I_{(k)}^* \cos[\theta_{(k)} + \varphi]}{i_{dc}^*} \tag{8.27}$$

According to (8.27), a proportional controller $P_3(z)$ is applied in the current loop, its proportional coefficient is donated by $k_{p3}$, $\varphi$ is the displacement angle of the grid current reference.

Moreover, a proportional-integral controller $PI_2(z)$ is applied in the voltage loop, which is represented as

**Table 8.2** Default system parameters

| Parameters | Value | Parameters | Value |
|---|---|---|---|
| $V_m$ | 156 V | $f_c$ | 5 Hz |
| $f_l$ | 50 Hz | $f_s$ | 20 kHz |
| $\omega_l$ | $100\pi$ rad/s | $T_s$ | 50 μs |
| $R_o$ | 10 Ω | $\xi$ | 0.707 |
| $R_f$ | 27 Ω | $k_{p1}$ | 18 |
| $L_{dc}$ | 5.5 mH | $k_{i1}$ | 514 |
| $L_1$ | 1.8 mH | $k_{p2}$ | $1 \times 10^{-5}$ |
| $L_f$ | 2 mH | $k_{i2}$ | $1 \times 10^{-3}$ |
| $C_f$ | 6.8 μF | $k_{p3}$ | 190 |
| $C_o$ | 30 μF | $i^* dc$ | 4 A |
| $C_a$ | 40 μF | $v^* com$ | 220 V |
| $C_b$ | 40 μF | $p_o$ | 160 W |

$$x_{17(k+1)} = x_{17(k)} + k_{i2}T_s e_{v(k)}$$
$$I_{(k)}^* = x_{17(k)} + \left(k_{p2} + k_{i2}T_s\right)e_{v(k)} \tag{8.28}$$

where $k_{p2}$ and $k_{i2}$ are proportional and integral coefficients of $PI_2(z)$ respectively. The default system parameters are listed in Table 8.2.

### 8.1.4 Modeling of the DHSS

The dynamics of the whole nonlinear time-periodic system is described as

$$x_{(k+1)} = F_{(k)}\left(x_{(k)}, u_{(k)}\right) \tag{8.29}$$

where $x \in \mathbf{R}^{17}$ and $u \in \mathbf{R}^3$ represent the state vector and the input vector respectively, and $F_{(k)}$ describes a nonlinear function of the mapping relationship from $x_{(k)}$ to $x_{(k+1)}$. Its linearized model is described as

$$\hat{x}_{(k+1)} = \left.\frac{\partial F_{(k)}}{\partial x_{(k)}}\right|_{\substack{x_{(k)} = \overline{x}_{(k)} \\ u_{(k)} = \overline{u}_{(k)}}} \hat{x}_{(k)} + \left.\frac{\partial F_{(k)}}{\partial u_{(k)}}\right|_{\substack{x_{(k)} = \overline{x}_{(k)} \\ u_{(k)} = \overline{u}_{(k)}}} \hat{u}_{(k)}$$

$$= G_{(k)}\hat{x}_{(k)} + H_{(k)}\hat{u}_{(k)} \tag{8.30}$$

Both the system matrix $G_{(k)}$ and the input matrix $H_{(k)}$ are $K$-periodic, determined by the steady-state trajectories $\overline{x}_{(k)}$ and $\overline{u}_{(k)}$. e.g., $G_{(k)}$ is donated by

$$\boldsymbol{G}_{(k)} = \begin{bmatrix} G_{1,1} & G_{1,2} & \cdots & G_{1,j} & \cdots & G_{1,17} \\ G_{2,1} & G_{2,2} & \cdots & G_{2,j} & \cdots & G_{2,17} \\ \vdots & \vdots & \ddots & \vdots & \ddots & \vdots \\ G_{i,1} & G_{i,2} & \cdots & G_{i,j} & \cdots & G_{i,17} \\ \vdots & \vdots & \ddots & \vdots & \ddots & \vdots \\ G_{17,1} & G_{17,2} & \cdots & G_{17,j} & \cdots & G_{17,17} \end{bmatrix}_{(k)} \tag{8.31}$$

where

$$G_{i,j(k)} = \frac{\partial x_{i(k+1)}}{\partial x_{j(k)}} \tag{8.32}$$

The small disturbance signals $\hat{x}_{(k)}$ and $\hat{u}_{(k)}$ are also $K$-periodic, which satisfies

$$\begin{cases} \hat{\boldsymbol{x}}_{(k)} = \hat{\boldsymbol{x}}_{(k+K)} \\ \hat{\boldsymbol{u}}_{(k)} = \hat{\boldsymbol{u}}_{(k+K)} \\ \boldsymbol{G}_{(k)} = \boldsymbol{G}_{(k+K)} \\ \boldsymbol{H}_{(k)} = \boldsymbol{H}_{(k+K)} \end{cases} \text{ and } K = f_s / f_l \tag{8.33}$$

Since $\boldsymbol{G}_{(k)}$ and $\boldsymbol{H}_{(k)}$ are periodic matrixes, their discrete Fourier transform (DFT) form are donated by

$$\begin{cases} \boldsymbol{G}_{(k)} = \sum_{n=-\infty}^{\infty} g_n e^{jn\omega_l k T_s} = \sum_{n=-\infty}^{\infty} g_n e^{jn\frac{2\pi}{K}k} \\ \boldsymbol{H}_{(k)} = \sum_{n=-\infty}^{\infty} h_n e^{jn\omega_l k T_s} = \sum_{n=-\infty}^{\infty} h_n e^{jn\frac{2\pi}{K}k} \end{cases} \tag{8.34}$$

where $g_n$ and $h_n$ are the $n$th Fourier coefficient matrixes of the system matrix and input matrix respectively. To analyze the frequency response of the system, letting $\hat{x}(t)$ and $\hat{u}(t)$ be modulated by the EMP signal, i.e.,

$$\begin{cases} \hat{\boldsymbol{x}}(t) = \sum_{n=-\infty}^{\infty} \hat{X}_n e^{jn\omega_l t} e^{st} \\ \hat{\boldsymbol{u}}(t) = \sum_{n=-\infty}^{\infty} \hat{U}_n e^{jn\omega_l t} e^{st} \end{cases} \tag{8.35}$$

where $X_n$ and $U_n$ are the $n$th Fourier coefficients of the state vector and the input vector respectively. With $t = kT_s$ and $z = e^{sT_s}$ this gives

$$\begin{cases} \hat{\boldsymbol{x}}_{(k)} = \sum_{n=-\infty}^{\infty} \hat{X}_n e^{jn\frac{2\pi}{K}k} e^{skT_s} = \sum_{n=-\infty}^{\infty} \hat{X}_n \left(z e^{jn\frac{2\pi}{K}}\right)^k \\ \hat{\boldsymbol{u}}_{(k)} = \sum_{n=-\infty}^{\infty} \hat{U}_n e^{jn\frac{2\pi}{K}k} e^{skT_s} = \sum_{n=-\infty}^{\infty} \hat{U}_n \left(z e^{jn\frac{2\pi}{K}}\right)^k \end{cases} \tag{8.36}$$

According to harmonic balance theory, substituting (8.34) and (8.36) into (8.30) gives

$$z \cdot e^{jl\frac{2\pi}{K}} \hat{X}_l = \sum_{m=-\infty}^{\infty} g_{l-m} \hat{X}_m + \sum_{m=-\infty}^{\infty} h_{l-m} \hat{U}_m \tag{8.37}$$

which describes the state equation of the $l$th harmonic of the system. Then, the DHSS model can be formed and expressed as

$$zN\hat{X} = \boldsymbol{\Gamma}\{\boldsymbol{G}\}\hat{X} + \boldsymbol{\Gamma}\{\boldsymbol{H}\}\hat{U} \tag{8.38}$$

where $N$ is a diagonal matrix, donated by

$$N = \text{diag}\left[\cdots, e^{-j2\frac{2\pi}{N}}, e^{-j\frac{2\pi}{N}}, 1, e^{j\frac{2\pi}{N}}, e^{j2\frac{2\pi}{N}}, \cdots\right] \tag{8.39}$$

and the Toeplitz transform of the matrixes $\boldsymbol{G}$ and $\boldsymbol{H}$ is denoted by $\boldsymbol{\Gamma}\{\boldsymbol{G}\}$ and $\boldsymbol{\Gamma}\{\boldsymbol{H}\}$ respectively, which maps the set of complex Fourier coefficients into a doubly infinite block Toeplitz matrix [2, 3], e.g., $\boldsymbol{\Gamma}\{\boldsymbol{G}\}$ is donated by

$$\boldsymbol{\Gamma}\{\boldsymbol{G}\} = \begin{bmatrix} \ddots & \vdots & \vdots & \vdots & \vdots & \vdots & \iddots \\ \cdots & \boldsymbol{G}_0 & \boldsymbol{G}_{-1} & \boldsymbol{G}_{-2} & \boldsymbol{G}_{-3} & \boldsymbol{G}_{-4} & \cdots \\ \cdots & \boldsymbol{G}_1 & \boldsymbol{G}_0 & \boldsymbol{G}_{-1} & \boldsymbol{G}_{-2} & \boldsymbol{G}_{-3} & \cdots \\ \cdots & \boldsymbol{G}_2 & \boldsymbol{G}_1 & \boldsymbol{G}_0 & \boldsymbol{G}_{-1} & \boldsymbol{G}_{-2} & \cdots \\ \cdots & \boldsymbol{G}_3 & \boldsymbol{G}_2 & \boldsymbol{G}_1 & \boldsymbol{G}_0 & \boldsymbol{G}_{-1} & \cdots \\ \cdots & \boldsymbol{G}_4 & \boldsymbol{G}_3 & \boldsymbol{G}_2 & \boldsymbol{G}_1 & \boldsymbol{G}_0 & \cdots \\ \iddots & \vdots & \vdots & \vdots & \vdots & \vdots & \ddots \end{bmatrix} \tag{8.40}$$

## 8.2 Stability Boundary Based on Eigenvalues Loci

### 8.2.1 Eigenvalues Loci Analysis

Multiplying $N^{-1}$ on both sides of (8.38) gives

$$z\hat{X} = N^{-1}\boldsymbol{\Gamma}\{\boldsymbol{G}\}\hat{X} + N^{-1}\boldsymbol{\Gamma}\{\boldsymbol{H}\}\hat{U} \tag{8.41}$$

then applying Z-inverse transformation to (8.41) gives

$$\hat{X}_{(k+1)} = N^{-1}\boldsymbol{\Gamma}\{\boldsymbol{G}\}\hat{X}_{(k)} + N^{-1}\boldsymbol{\Gamma}\{\boldsymbol{H}\}\hat{U}_{(k)} \tag{8.42}$$

where $N^{-1}\boldsymbol{\Gamma}\{\boldsymbol{G}\}$ is the Jacobian matrix of the DHSS model, donated by

$$\boldsymbol{J} = N^{-1}\boldsymbol{\Gamma}\{\boldsymbol{G}\} \tag{8.43}$$

The stability of the system can be analyzed by the eigenvalues of the Jacobian matrix $J$ in the complex plane. The convergence and divergence of the state vector is determined by the amplitude of the corresponding eigenvalue. If its amplitude is greater than one, the corresponding state variable is divergent. Moreover, the oscillation frequency of the state variable is determined by the argument (angle between vector and positive real axis) of the corresponding eigenvalue. The smaller the argument (the eigenvalue is closer to the positive real axis), the lower the oscillation frequency.

The Jacobian matrix $J$ is an infinite dimensional matrix theoretically, truncation is usually necessary for practical analysis, and the truncation order $h$ depends on the required accuracy. Figure 8.5 shows the distribution of eigenvalues at different truncation orders. As seen, the number of eigenvalues (i.e., dimension of Jacobian matrix) increases with the increase of truncation order $h$. The difference between the center frequency and the edge frequency is $hf_1$, the larger the truncation order $h$, the wider the frequency range involved. Moreover, with the increase of truncation order $h$, the shift of center frequency becomes smaller. When $h > 3$, the center frequency hardly shifts, which means that the truncation error decreases as the truncation order $h$ increases. Unfortunately, an explicit formula for error cannot be provided. Considering the accuracy and complexity comprehensively, the truncation order $h$ is set to 7 in the following work.

To verify the proposed method, when $L_{dc}$ is set as 5.5 mH, the eigenvalues loci under different $k_{p3}$ are shown in Figs. 8.6 and 8.7. As shown in Fig. 8.6b, the parameter $k_{p3}$ is varied from 60 to 120. When $k_{p3}$ is set as 100 or 120, all eigenvalues are within the unit circle, indicating the system is stable. When $k_{p3}$ is set as 60, some eigenvalues move out of the unit circle, indicating the system is unstable. When $k_{p3}$ is set as 80, some eigenvalues are very close to the unit circle and the central frequency $f_{sol}$ is 601 Hz. Once these eigenvalues move out of the unit circle, the system will generate oscillations near 601 Hz. With the decrease of $k_{p3}$, these eigenvalues are farther from the unit circle,

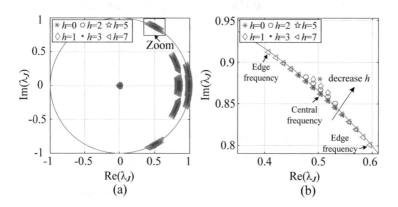

**Fig. 8.5** Comparative analysis of eigenvalues with different truncation orders when $k_{p3} = 216.2$. **a** Overall eigenvalues loci. **b** Zoom of (**a**)

the oscillation phenomenon will be stronger, and the system will even enter a chaotic state.

Similarly, as shown in Fig. 8.7b, when $k_{p3}$ is set as 205 to 215, all eigenvalues are within the unit circle. But when $k_{p3}$ is set as 220, some eigenvalues move out of the unit circle and the central frequency $f_{so2}$ is 3332 Hz, indicating the system will generate oscillation near 3332 Hz.

Figure 8.8 shows the eigenvalues loci under different $L_{dc}$, when $L_{dc}$ is set as 56 mH, all eigenvalues are within the unit circle. But when $L_{dc}$ is set as 4.5 mH, some eigenvalues move out of the unit circle and the central frequency $f_{so3}$ is 3389 Hz, indicating the system will generate oscillation near 3389 Hz.

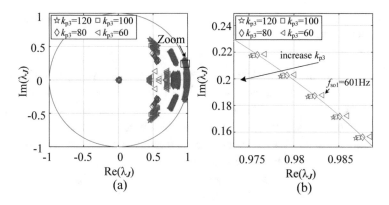

**Fig. 8.6**  Eigenvalues loci under $L_{dc} = 5.5$ mH when $k_{p3}$ is between 60 and 120. **a** Overall eigenvalues loci. **b** Zoom of (**a**)

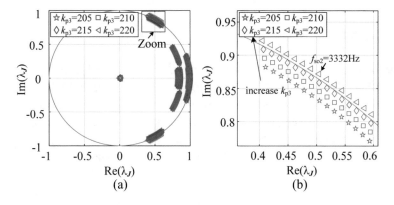

**Fig. 8.7**  Eigenvalues loci under $L_{dc} = 5.5$ mH when $k_{p3}$ is between 185 and 230. **a** Overall eigenvalues loci. **b** Zoom of (**a**)

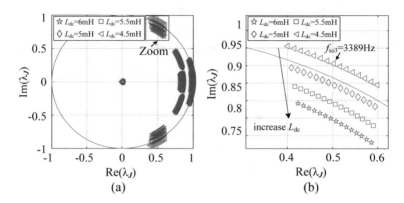

**Fig. 8.8**  Eigenvalues loci under $k_{p3} = 190$ when $L_{dc}$ is between 4.5 and 6 mH. **a** Overall eigenvalues loci. **b** Zoom of (**a**)

Figure 8.9 shows the eigenvalues loci under different $k_{p2}$. All eigenvalues are within the unit circle when $k_{p2}$ is set as $1 \times 10^{-4.2}$ to $1 \times 10^{-3.3}$, but the several eigenvalues of edge frequency move out of the unit circle when $k_{p2}$ is set as $1 \times 10^{-3.6}$, indicating that harmonics of corresponding order are unstable. When $k_{p2}$ is set as $1 \times 10^{-3.3}$, the eigenvalues near $f_{so4}$ and $f_{so5}$ move out of the unit circle, indicating that the system enters multi frequency oscillation even a chaotic state.

The several main oscillation frequencies of system are listed in Table 8.3. Under the system parameters in Table 8.2, the system is easy to generate the oscillations near $f_{so1}$ and $f_{so2}$ by adjusting $k_{p3}$, to generate the oscillations near $f_{so3}$ by adjusting $L_{dc}$, and to generate the oscillations near $f_{so4}$ and $f_{so5}$ by adjusting $k_{p2}$. It can be seen that the oscillation frequencies are concentrated around 300, 600 and 3300 Hz.

### 8.2.2  Stability Boundary

In order to further study the relationship among $k_{p3}$, $\lg(k_{p2})$ and $L_{dc}$ and system stability, the three-dimensional (3D) stability boundary is rendered in Fig. 8.10, where the region between the two planes is stable. It clearly shows the stability boundary of $k_{p3}$ and $L_{dc}$ exhibits a linear relationship, illustrating a fact that a larger DC inductance $L_{dc}$ will bring a wider stable operation range. However, the stability boundary of $\lg(k_{p2})$ and $k_{p3}$ exhibits a nonlinear relationship. Either too large or too small $k_{p2}$ will compress the stable region of $k_{p3}$.

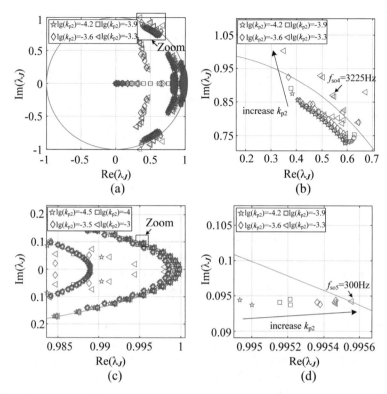

**Fig. 8.9** Eigenvalues loci under $k_{p3} = 190$ and $L_{dc} = 5.5$ mH when $k_{p2}$ is between $1 \times 10^{-4.2}$ and $1 \times 10^{-3.3}$. **a** Overall eigenvalues loci. **b** Zoom of (**a**). **c** Local eigenvalues loci. **d** Zoom of (**c**)

**Table 8.3** Main oscillation frequencies of system

| Symbol | Eigenvalues $\lambda$ | $|\lambda|$ | Frequency (Hz) |
|---|---|---|---|
| $f_{so1}$ | $0.983105 \pm j0.187726$ | 1.000867 | 601 |
| $f_{so2}$ | $0.503868 \pm j0.871776$ | 1.006914 | 3332 |
| $f_{so3}$ | $0.501924 \pm j0.905881$ | 1.035638 | 3389 |
| $f_{so4}$ | $0.542452 \pm j0.870033$ | 1.025286 | 3225 |
| $f_{so5}$ | $0.995554 \pm j0.094222$ | 1.000002 | 300 |

## 8.3   Experimental Verifications

An experimental prototype is built to verify the stability boundary of system parameters. The experimental waveforms under different coefficient $k_{p3}$ are shown in Fig. 8.11, and other parameters remain the same as Table 8.2. It can be observed that no unstable oscillations appear in the output current $i_{dc}$, output voltage $v_o$ and grid current $i_g$ when $k_{p3}$ is

**Fig. 8.10** 3D stability
boundary among $k_{p3}$, $\lg(k_{p2})$
and $L_{dc}$

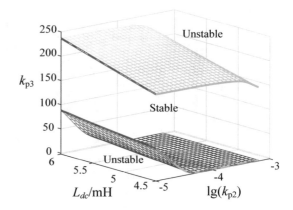

set as 85 or 210, which means the system is stable. However, when $k_{p3}$ is set as 75, the
fast Fourier transform (FFT) of $i_g$ shows that the system generate oscillation near $f_{so1}$.
Moreover, when $k_{p3}$ is set as 220, it can be seen that the obvious oscillations near $f_{so2}$
appear in the system. Thus, the experimental results satisfy the analysis of eigenvalues
loci in Figs. 8.6 and 8.7.

The experimental waveforms under different $L_{dc}$ are shown in Fig. 8.12, and other
parameters remain the same as Table 8.2. As seen, when $L_{dc}$ is set as 5mH, there are no
obvious oscillations existing in the output current $i_{dc}$, output voltage $v_o$ and grid current
$i_g$. However, when $L_{dc}$ is set as 4.5 mH, the oscillations near $f_{so3}$ appear in the system
according to the FFT of $i_g$. The experimental results satisfy the analysis of the eigenvalues
loci in Fig. 8.8.

Similarly, the experimental waveforms under different $k_{p2}$ are shown in Fig. 8.13. As
seen, there are no obvious oscillations existing in the system when $k_{p2}$ is set as $1 \times 10^{-4}$,
but strong oscillations near $f_{so4}$ and $f_{so5}$ appear in the grid current $i_g$ when $k_{p2}$ is set as
$1 \times 10^{-3.3}$. The experimental results satisfy the analysis of eigenvalues loci in Fig. 8.9.
According to the above FFT results of $i_g$, all experimental results are consistent with the
oscillation frequencies in Table 8.3.

The comparison result of the stability boundary is illustrated in Fig. 8.14. It can be
seen that stable experimental results lie within the stable region, the unstable experimental
results lie within the unstable region, thus the accuracy of the stability boundary is verified
by the experimental results. Existing errors may come from the following aspects: (1) the
truncation error of Toeplitz matrix, (2) the limited step of numerical analysis, (3) The
saturation effect of inductance is not considered in the DHSS model, thus the inductance
is not an absolute constant.

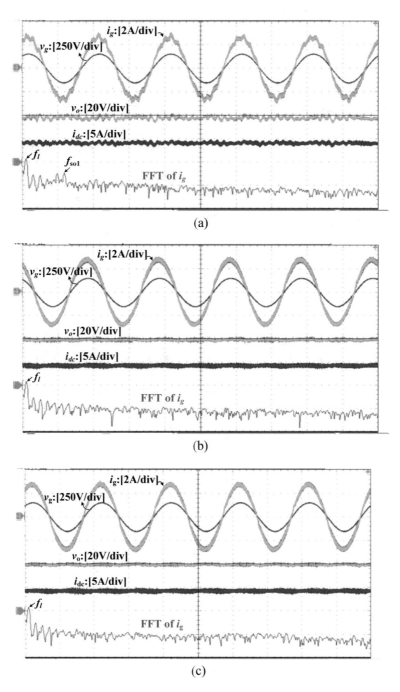

**Fig. 8.11** Experimental waveforms under different $k_{p3}$ when $L_{dc} = 5.5$ mH and $k_{p2} = 1 \times 10^{-5}$. **a** $k_{p3} = 75$. **b** $k_{p3} = 85$. **c** $k_{p3} = 210$. **d** $k_{p3} = 220$

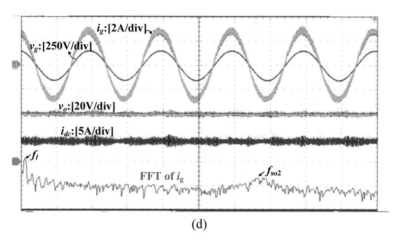

(d)

**Fig. 8.11**  (continued)

## 8.4  Conclusion

In this chapter, a DHSS modeling method is proposed to analyze the stability of a high dimensional single-phase current source power decoupling converter with digital controller. Through the DSS model, the system matrix during a power frequency period are obtained, and then a Toeplitz matrix is obtained by DFT. Further, the system stability can be analyzed by the Jacobian matrix eigenvalues of DHSS model. The proposed model can be regarded as a combination of DSS model and HSS model. It remains the accuracy of DSS model and the autonomous nature of HSS model. The whole modeling process is based on computer-aided numerical analysis, which avoids mass manual calculations and solves the modeling challenge from complex and non-analytical power electronic converter. Finally, the experimental results validate the accuracy of DHSS model.

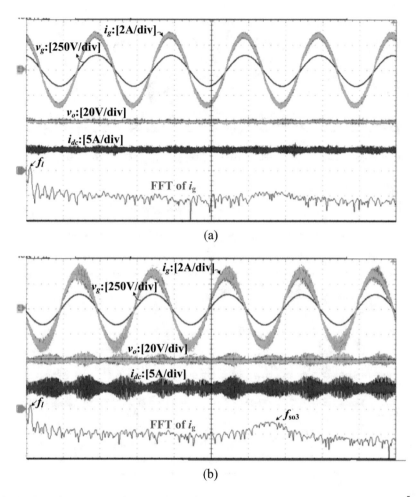

**Fig. 8.12** Experimental waveforms under different $L_{dc}$ when $k_{p3} = 190$ and $k_{p2} = 1 \times 10^{-5}$. **a** $L_{dc}$ = 5 mH. **b** $L_{dc}$ = 4.5 mH

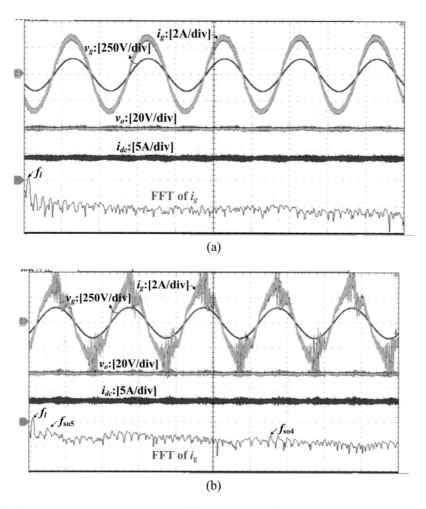

(a)

(b)

**Fig. 8.13** Experimental waveforms under different $k_{p2}$ when $L_{dc} = 5.5$ mH and $k_{p3} = 190$. **a** $k_{p2}$ $= 1 \times 10^{-4}$. **b** $k_{p2} = 1 \times 10^{-3.3}$

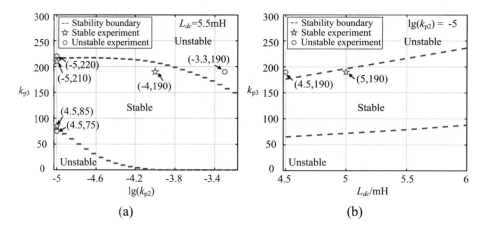

**Fig. 8.14** Comparison of stability boundary and experimental results when $k_{p2} = 1 \times 10^{-4}$ (**a**), and $k_{p2} = 1 \times 10^{-3.3}$ (**b**)

## Bibliography

1. Zhang, C., Molinas, M., Føyen, S., Suul, J. A., & Isobe, T. (2020). Harmonic-domain SISO equivalent impedance modeling and stability analysis of a single-phase grid-connected VSC. *IEEE Transactions on Power Electronics, 35*(9), 9770–9783.
2. Mllerstedt, E. (2000). *Dynamic analysis of harmonics in electrical systems.*
3. Lin, J., Su, M., Sun, Y., Yang, D., & Xie, S. (2022). Recursive SISO impedance modeling of single-phase voltage source rectifiers. *IEEE Transactions on Power Electronics, 37*(2), 1296–1309.
4. Wereley, N. M. (1991). Analysis and control of linear periodically time varying systems. Dissertation Department of Aeronautics Astronautics, Massachusetts Institute of Technology.
5. Kuznetsov, Y. A. (2004). Elements of applied bifurcation theory. *Applied Mathematical Sciences 288*(2), 715–730.
6. Liu, Y., Tang, S., Wang, H., Ning, G., & Xiong, W. (2021). Independent power decoupling method using minimum switch devices for single-phase current source converters. *Journal of Power Electronics, 21*(9), 1383–1394.